THE APPEARANCE OF THE FORM.
N. J. HABRAKEN

Routledge Revivals

The Appearance of the Form

Originally published in 1985, this book explores, in four interwoven essays, the many ways human life and built form interact and the place that professional designing takes in this interaction. Together, the essays touch on a number of ideas: the idea that our position in space relative to the thing we are designing determines the methods we apply when designing it; the idea that designing is about making proposals, and is therefore a social act first of all; and the idea that agreements, consensus and above all conventions shape the act of designing things independent of their creative qualities.

The Appearance of the Form

Four Essays on the Position Designing takes between People and Things

by N.J. Habraken

First published in 1985
by Awater Press
Second Edition published in English 1988
by Awater Press

This edition first published in 2019 by Routledge
2 Park Square, Milton Park, Abingdon, Oxon, OX14 4RN

and by Routledge
605 Third Avenue, New York, NY 10017

Routledge is an imprint of the Taylor & Francis Group, an informa business

© 1985 Awater Press
© 2019 N.J. Habraken

All rights reserved. No part of this book may be reprinted or reproduced or utilised in any form or by any electronic, mechanical, or other means, now known or hereafter invented, including photocopying and recording, or in any information storage or retrieval system, without permission in writing from the publishers.

Notice:
Product or corporate names may be trademarks or registered trademarks and are used only for identification and explanation without intent to infringe.

Publisher's Note
The publisher has gone to great lengths to ensure the quality of this reprint but points out that some imperfections in the original copies may be apparent.

Disclaimer
The publisher has made every effort to trace copyright holders and welcomes correspondence from those they have been unable to contact.

ISBN 13: 978-0-367-85773-8 (hbk)
ISBN 13: 978-1-003-01503-1 (ebk)
ISBN 13: 978-0-367-85780-6 (pbk)

N. John Habraken

THE APPEARANCE OF THE FORM

Four Essays on the Position Designing takes between People and Things

First published 1985
Second edition 1988
This digital edition published in 2014 by
Awater Press
Austin, Texas, USA

Copyright © Awater Press

All rights reserved.
No part of this book may be reproduced in any form or by any electronic or mechanical means including information storage and retrieval systems, without permission in writing from the publisher. The only exception is by a reviewer, who may quote short excerpts in a review.

ISBN-978-0-9913019-0-4

Other English language books by John Habraken:

Supports: An Alternative to Mass Housing, 1961 (in Dutch). First English edition 1972; Second English edition, Edited by Jonathan Teicher, Urban International Press U.K., 2011.

Three R's for Housing. originally published in Forum, Vol xx, Nr.1. First printed edition by Scheltma &Holkema, 1970.

Variations: the Systematic Design of Supports, with J.T.Boekholt, A.P.Thijssen, P.J.M.Dinjens. Dept of Architecture, MIT, 1975.

The Grunsfeld Variations: A Report on the Thematic Development of an Urban Tissue, With José Aldrete Haas, Renee Chow, Tom Hille, Paula Krugmeier, Martha Lampkin, Anthony Mallows, Andrés Mignucci, Yutaka Takase, Kimberly Weller, Toshihito Yokouchi. Department of Architecture MIT. 1981.

Transformations of the Site, 3d revised edition, Awater Press, 1988.

The Structure of the Ordinary, Form and Control in the Built Environment. Edited by Jonathan Teicher. MIT Press, 1998.

Palladio's Children, Seven essays on everyday environment and the architect, Edited by Jonathan Teicher, Taylor and Francis, 1st edition 2005.

Conversations with Form. N.J.Habraken, Andrés Mignucci and Jonathan Teicher, Routledge, 2014.

For downloadable papers and editions, including in other languages, see: www.habraken.org

de·sih 'nlcnt; *n.* the act of making unsightly; disfigurement. [Rare.]
de·sign' (-zin'). *v.t.* i de signed, *pt.*, *pp.*; designing, *ppr.* [OFr. *desrgner* L. *desig rare*, to mark out, to define; *de*, out, from, and *signore*, to mark, from *sign um*, a mark, sign.]
1. to plan and delineate by drawing the outline or figure of; to sketch, as in painting and other works of art, as for a pattern or model.
2. to contrive; to project with an end in view; to form in idea, as a scheme.

Ask of politicians the end for which laws were originally *designed*. -Burke.

3. to intend or set apart for some purpose.
One of those places was *designed* by the old man to his son. -Clarendon.
4. to purpose; to intend; as, a man *designs* to write an essay, or to study law.
5. to decide upon and outline the main features of; to plan.
6. to indicate. [Rare.]
Syn.-intend, plan, propose, purpose, sketch.
d ,ign', *v. i.* 1. to make designs.
2. to make original plans, sketches, patterns, etc.; as, she *designs* for a coat manu- f act urer.
de·{lign', *n.* 1. a plan; scheme; project.
2. **purpose; intention; aim.**
3. a thing planned for or outcome aimed at.
4. a working out by plan; as, do we find a *design* in history?
S. [pl.] a secret or sinister scheme (often with *on* or *upon);* as, he has *designs* on her property.
6. a plan or sketch to work from; a pattern; as, a *design* for a house.
7. the art of making designs or patterns.
8. the arrangement of parts, details, form, color, etc., especially so as to produce a complete and **artistic unit; artistic invention;** as, the *dtsig t* of a rug.
9. a finished artistic work.
d ign'il·ble, *a.* capable of being designed or marked out ; distinguishable.

Dedicated

To the memory

of

Piet Sanders

and

Ida Sanders

in recognition

of their help and encouragement

through many years

Say it,
no
ideas
but
in
things.

(William Carlos Williams, *Paterson*)

PREFACE 12

one: SHARING 14

THE SOURCE 15
CONVENTIONAL FORM 19
THE SHARED IMAGE: BUILDING 22
THE SHARED IMAGE: MANUFACTURING 25
BORROWING INFORMAL TECHNOLOGY 28
BORROWING FORMAL TECHNOLOGY 31
LIVING AND MAKING 35
TYPE 39
RANGE OF VARIATION 42
DESCRIBING 47
NAMING 50
THE LARGER CONTEXT 53
THE SHAPE OF PUBLIC SPACE 56
SYSTEMS 60
KINDS OF SYSTEMS 63
THE DESIGNER 67

two: DESIGNING 69

THE APPEARANCE 70
THE TASK 71
THE FORM 73
CONVENTION 75
CREATIVITY 79
ROLE AND TASK 81
STRING AND SEQUENCE OF STEPS 84

EXAMPLES 89
BALANCE 93
CONSTRAINTS AND FORMS 95
EXPLICIT AND IMPLICIT CONSTRAINTS 99
SOLUTION SPACE 103
A SIMILAR RELATION 108
THE RIGHTWARD MOVE 113
THE LEFTWARD MOVE 115

three: SEEING 119

THEMES AND SYSTEMS 120
SOCIAL BONDAGE 123
FORMS AND WORDS 127
IDENTITY 130
BACK TO THE MODEL 132
HOUSE AND MACHINE 136
CAPACITY AND FUNCTION 140
EXPLORING CAPACITY 143
DESIGN ATTITUDES 147
HIERARCHY 151
INTERVENTION 156
DESIGNING WITH HIERARCHIES 160
CONSTRAINTS FROM ABOVE 164
CONSTRAINTS FROM BELOW 167
LOOKING THE OTHER WAY 170
THE MODEL EXPANDED 173

four: CONTROLLING
176

WHOLES	177
THINGS AND PEOPLE	180
ATTACHMENTS	182
THE USERS WORLD	186
THE AUTONOMY OF MAKING	192
BORROWED CONTROL	195
HIERARCHY IN FORM	198
HIERARCHY IN SPACE; TERRITORY	203
TERRITORIAL HIERARCHY	207
LEARNING FROM SETTLEMENT FORMS	211
NEGOTIATION	213
LEARNING FROM SETTLEMENT FORMS II	219
DIFFERENCES	223
MATCHING HIERARCHIES I	226
MATCHING HIERARCHIES II	229
LAST LOOK AT THE MODEL	232
THE SOURCE	235
ILLUSTRATIONS	238
ENDNOTES	240

PREFACE

These four essays were written when I taught my *Thematic Design* course at MIT (1982-88). This was an exercise-based course in our Second Masters in Architectural Studies (SMArchS) degree program for students who already had a professional degree in architectural design. My course recognized the habitual involvement of contemporary professional designers in the workings of everyday built environment and focused on design skills needed to help cultivate it. Its exercises had to do with transformations over time, typology, the hierarchical structure of complex urban fabrics, and the capacity of architectural and urban spaces to accommodate varied and variable uses.

This approach obviously encouraged observation of historic urban fabrics - extremely complex entities that had endured for centuries as a kind of living organisms - in an attempt to learn from them what general principles they had in common that might be helpful in our work today.

The methodological approach of my course simply assumed that the professional designer had a role to play in the development and wellbeing of contemporary everyday environment. But I knew that historic urban fabrics of great complexity and beauty had come about without involvement of professional design as we know it today. This, in turn, raised the question as to how, in general, the many things that we make come into the world. Apparently, humanity has the innate ability, deeply rooted in its social body, to bring forth sophisticated and complex physical objects without involvement of professional designers as we know them today. While designing, as we know it, is a relatively recent professional phenomenon, how is it embedded in that fecund social context? In other words, how is designing as we know it part of the age-old creative process that brings forth the many things that we live with? The essays in this book are explorations of the way artifacts appear in human society, each one of them looking for conditions and circumstances from a somewhat different angle.

The questions that triggered these attempts remain, I believe, as relevant today as they were when this book was first written.

one: SHARING

THE SOURCE

We have always made things to live with. We do not eat, cook, dress, sleep, move, or dwell without artifacts to assist us. Nor can we cultivate the earth or exploit its resources without them. In this culture of people and things, the designer has appeared as someone who produces a plan for what is to be made. Designing is one of several activities we engage in to supply ourselves with the artifacts we want around us. The term 'designing' in the way we understand it today, as the making of a plan to inform making, only came into our language with modern times. Later, people who were called designers became a professional species distinct from the makers of things. Making is definitely a much older activity, one that already long ago became a professional occupation separating the user from the maker. Of course, acts of designing took place very early in history. We may safely assume that scale models, drawings or diagrams of some kind have preceded the building of many of the monumental buildings of the past. Indeed, architecture may well be the field where designing first became a formal activity. Yet, apart from the erection of structures like the palace, the temple, the castle or the mausoleum, building was an activity that did not know designing and indeed, we have no reason to believe that monumental projects always were executed following documents we now call 'designs.'

We know that most towns in history just grew incrementally following paths that in turn found their way among riverbanks and slopes. It is a long way from the organic growth of settlements via preconceived lay-outs to the professional town design that we know today. David Friedman, discussing early renaissance Florentine towns, identifies a point where professional design emerges in town building: "The presence of geometry would, also, establish the new town plans in the realm of fine arts because it testifies that the authors of the plans were professional designers. Many medieval towns were laid out by simple surveyors, following the instructions of city officials... These plans invariably lack a geometric base. Simple repetitive street patterns and mimetic dependence on earlier projects gives these projects shape." [1]

Thus, while the activity we call designing must have been present for a long time in many cultures, it was nevertheless an exceptional one and its products were few. We can safely say that most things produced were not designed. They were created with deliberation and intentions, yes: but not via the product we call 'a design' or 'a plan' intended to instruct their making; much less with the help of individuals we would call 'designers.' Until this day, forms of great complexity - buildings, ships, tools, utensils, and art works - do come about without a professional designer's efforts, indeed, as I hope to show, often without much in the way of designing at all.

Each period has its ways in which people depend on artifacts and in which artifacts come into existence; each period has its ways in which production is shaped by social structure and society organizes itself around production. In the beginning of modern times a new trend can be noted, this time to yield a large place for the kind of making that does not produce the artifact itself but a plan, a proposal about what the artifact should be.

We still make new things without designing. The new way of working did not replace old ways but was added to them. Nevertheless, designing is so pervasive an activity today that the professional designer, who is the product of this new trend, may be excused for believing that all artifacts have been designed. Therefor it is good to remind ourselves that this is not the case. Only under certain circumstances does designing precede the making of things.

If designing is to instruct making, it must produce something that represents what is to be made. The designer's world is one of representations. We describe the object to be produced in a hundred ways, by means of drawings, words, and diagrams, in print, by scale models or on computer screens. In our descriptions we use a variety of formal languages tied to as many specializations. In aid of production of such diverse things as buildings, machines, automobiles, or ships there are often specifications of all parts and drawings of all joints. All dimensions are stated; calculations and evaluations of numerous aspects are on record. The thing to be made appears among us in many ways before it is actually produced. For those who live by designing the representation is the substance and what cannot be documented and described escapes the designer's domain.

If designing is the result of circumstances that trigger the making of representations, we would do well to observe those circumstances. We will learn more about a plant when we ask ourselves what kind of climate it needs and in what soil it grows best. What, then, makes a design come about? What are the conditions within which designing becomes a recognizable and necessary activity? What can be done by designing that cannot be achieved without it? What, on the other hand, cannot be done by designing?

Such questions suggest an investigation that is far beyond what I dare to engage in. But perhaps a contribution can be made by looking into a few examples of cases where designing is not yet emancipated from making. What is a world like in which designing is not a formal activity? There is no reason to presume that its artifacts are less complex or the social relations they are connected to more primitive than ours before we have found evidence to the contrary.

The practice of designing that we are familiar with today grew from this older mode of making. It is what our work as professional designers must be measured against: if we knew more about the source designing springs from, we might be in a better position to explain what particular contribution we, as professional designers, can make to the cohabitation of artifacts and people.

CONVENTIONAL FORM

Edwin Tappan Adney was born in 1868 and lived eighty-two years. All his adult life he studied the bark canoe as it was still used by the North American Indians of the lake region, making sketches and models of the various types in use and collecting information about the way they were made. This Robert McPhee tells us in his book on "The Survival of the Bark Canoe." He reports on the process of making the canoe in such detail that no reader can fail to be impressed by the technical genius invested in this object. From the stripping of the bark from a living tree to the forming of the skin and the careful building of the frame inside it, a minimum of tools was used to arrive at a product of great sophistication developed with superior intelligence. [2]

McPhee does not go beyond the building process. We can only speculate how the bark canoe emerged in the Indian culture. We may assume that, over time, improvements came about under the eyes of alert individuals. Perhaps the entire development of the canoe, up to the time that it became extinct, along with the culture that produced it, was slow. More primitive prototypes must have preceded the models we know from the accounts of Adney and others. It may have been a development in fits and starts, but then again it may have occurred in a relatively short time, after which the various types remained relatively steady. It may have been that some people in those days squatted on a flat piece of soil to sketch in it or made a scale model discussing a way to improve on a more primitive variant. Most likely, experiments were conducted and trial runs made. Passionate debates must have taken place about alternative details. We do not know.

But whatever assumptions may be correct, we can be assured that the sophisticated technology embedded in the canoe was the result of immediate interaction between people and artifact, without many representations in between, if at all.

It may well have been that a certain role distribution took place, as is the case with the hollowed tree canoes of the Papuas where some families did canoes and others did pottery.[3] McPhee tells us that bark canoes were built on fixed places, specially prepared to accommodate the work and in use for generations. Even when we assume there were individuals who knew how to build canoes and others who did not, we still have a situation where using and making were barely separated. Almost certainly there was no separation of designing from making. Of course, there must have been individuals who spent a good deal of time thinking about a better way to do a detail. It is clear from the product as we know it that the North American Indians had their share of technical talent. A product like the bark canoe can only have come about the way all cultures achieve high quality artifacts, by the continuous application of intelligence and creativity.

The bark canoe, I suggest, underwent continuous transformations before it reached the shape and make we now know. This could happen because many minds could share the form over a long time. The canoes where known in a social body; people lived with them and while these artifacts were among them they were kept in a state of continuous improvement towards perfection. This presence of an image as a collective knowledge in the minds of many, continuously reinforced by the presence of the artifacts themselves is what constitutes 'Conventional Form.' It is a form the image of which we all share. Conventional form is self-evident form.

THE SHARED IMAGE: BUILDING

As a young architect with little experience I was hired to oversee the execution of a large commercial 'fast track' project. I had the good fortune to work on the job with a seasoned daily supervisor who, while we climbed the scaffolding and navigated the muddy site, taught me much of what one never learns at school about the practice of building. He handled the complexities of reinforced concrete, steel and brick construction, heating and ventilation systems, and electric circuits. He inspected carpentry, the laying of bricks, the moving of earth, the driving of piles, the application of paint, marble, and tiles. He dealt not only with the workers on site, but also with manufacturers, contractors, and the occasional bureaucrat, doing it all with gusto and good humor.

When we could relax and have a cup of coffee, he liked to talk of his younger years when he was a small contractor building townhouses. He claimed that in those days, things were much better organized. "We had only one drawing" he said, "and that one, like those you make today, had to be improvised upon if one wanted to stay out of trouble. Measurement was no problem. All dimensions were expressed in the module of a brick. Anyone on the job could check a dimension. Windows and door frames were indicated by type and their dimensions specified in brick masonry modules - so many courses high, and so many headers wide. This was enough for the carpenter to get to work. Profiles and joints were common knowledge and needed no specification. All materials came on site exactly when needed and were installed immediately. Precut lumber for floors and the roof was loaded on a horse drawn wagon in reverse sequence so that it came out the correct way. You brought it right where it belonged in the building. Everybody knew every detail, thus there was continuous inspection; you could not claim ignorance. We did an entire town house of three floors, a cellar and an attic with a handful of men in a couple of weeks." In this way he would go on, describing in detail how the building was erected by what seemed to me a well-rehearsed choreography of men and materials in the course of which things fell into place effortlessly.

Couched in obvious exaggerations the message was clear: In those days the building to be built was in the heads of those doing the work on the site. That is not to say that each townhouse was a copy of the previous one, but all were variations on a familiar type. What particular variant had to be made was shown on the one drawing done by an architect who "only came back to instruct the plasterers on ceiling decorations and the painters on colors."

I had the good fortune to learn about the building of conventional form by exposure to oral history.

Years later I heard of a similar experience when a friend and colleague told me about his apprenticeship. He was the son of a farmer who could not afford to have him attend more than primary school. Because he seemed bright, he was made to work for the local carpenter and general contractor when he was twelve years old. One day a farmer walked into the workshop. After the usual exchange about the weather and the crops the carpenter asked what he could do. "We need a new house for the priest" answered the farmer who spoke on behalf of the congregation of the parish. "That's good, where is it to be built?" asked the carpenter. The location was explained. There followed some discussion about how the priest needed, unlike other folk, a room to study and receive visitors. Then the carpenter said: "I figure I can begin late next week." Thus, it was agreed and, although the issue of the extra room came up again later, this was about as close as the carpenter and his client got to designing.

THE SHARED IMAGE: MANUFACTURING

In 1982, the writer Michael Lenehan wrote an article about the making of a Steinway piano. For months he followed the genesis of the Steinway Grand K2571 from the selection of the wood down to the tuning of the finished instrument.[4]

His account describes the work of a variety of craftsmen, each a master in his own right, working with different materials and different tools on different parts of the instrument. They were all perfectionists and proud of their work. In the end he asked to see the design of the Grand and to his astonishment he was told that there was no such thing. There had never been one for any of the firm's models. The closest he could get was that drawings had been made after the fact to show important aspects of the available model.

It became clear that no design was needed for the coordination of the work of the various craftsmen who worked on it. Everyone knew what to do and, most importantly, what others were doing. The use of full-scale templates and measurement sticks proved sufficient to provide the necessary coordination and instruction.

The Grand, like all other types done by the firm, had evolved on the shop floor from previous models by trial and error. Nobody seemed to remember exactly how. Everything eventually related to the "scale stick" that showed the distances between the places where the hammers must strike the strings. From the scale stick, Lenehan was told, one could build a piano. The fact that all the workers knew the whole process, although each was a specialist for only part of it, made for a flexible operation. Modifications of details could occur by means of consultation among those involved. And anyone could determine who was involved in any problem that might come up. That does not mean, of course, that there was no hierarchy in the organization or that management did not play an important role.

In this way, the Steinway pianos went, over time, through a process of continuous, but gradual improvement that made them the superb instruments we know today.

We find here another process based on a shared image, aspects of which are similar to the examples given earlier. Two conditions seem to be important in all cases: That the composition of the whole is shared knowledge, and that each person is in full control of a specific task. These two things go together and depend on one another.

The success of the Steinway factory is particularly related to the social structure behind it. The workers were immigrants and children of immigrants with strong ties to the company for several generations. The company was their world. It took care of them and, in many cases, had been the reason for them to leave their native lands. Expertise grew in the factory and remained there, often passed on from father to son, to contribute to the improvement of the product. There was, in short, a social body that produced the Steinway piano. The form held these people together as much as it was their product. This indicates a third condition that seems even more important for the successful cultivation of conventional form: There must be time for knowledge of the form to take root in the social body that shares it.

BORROWING INFORMAL TECHNOLOGY

In the early forties the Egyptian architect Hassan Fathy decided that the mud-brick vaults and domes that had been erected in the Nile valley in the past were still worth building today. He correctly argued that the material was not only cheap, but that the product was superior to the flat roofed concrete architecture advocated in his time. It provided better insulation from the heat and was far more attractive. The domes and vaults could be built quickly, he presumed, with little or no need of scaffolding. He knew that they were built in the past without use of formwork and tried to re-invent the process but failed in a number of experiments. Finally, during a visit to the upper Nile region of Nubia he found these forms still very much in use, several examples having been newly erected. Upon his inquiry as to how he could meet someone who knew how to make these structures the reaction was one of amusement. Everybody knew how to do that! Practical know-how, once you have it, is the easiest thing in the world. [5]

Fathy's report on his many disappointments trying to get this traditional technology generally accepted again is a moving and sad story. It also is an excellent lesson for us to learn how making, using, and designing are forms of control and therefore make technology dependent on relations among people. For a form to be part of a culture it must be congruent with the interests that are peculiar to the day and age. In addition, efficiency is always related to a body of actors.

Fathy ran headlong into the opposition of bureaucrats, fellow architects and engineers who saw no merit in a technology they were not familiar with and that, they feared, did not need them and was alien to the modern forms they collectively held in mind. He also collided with the expectations of the users who saw mud-brick structures as remnants of a past they did not want to be part of anymore.

Fathy himself did not advocate that the domes and vaults be built by the people themselves but wanted to make an architecture that he judged to be best for them. His work was indeed based on a deep understanding of the performance of traditional local built environment. However, he was the professional architect, staying in full control of the building's creation to hand it to the people for their use.

The villages that Fathy designed were not completed. What did get built was gradually transformed by its occupants, using available materials like concrete blocks, baked brick, wood and corrugated metal, trying to accommodate an evolving lifestyle. Fathy did successfully apply the mud-brick vaulting technique in a few beautiful houses for well-educated clients who recognized his talents and shared his love for those organic shapes and the dignified, cool spaces with tempered light inside.

The lesson for us is not that a form will not stay for reasons of beauty, cost efficiency, climate, or geography but that each time again a match must be found between people and forms, and only when it is found both will prosper. Perhaps the mud-brick vault will, eventually, have its day in modern society.

A future generation of architects, engineers and users may be more interested in its qualities. Status and meaning may once more be attached to humble traditional materials. But the mudbrick vault represents a way of working that was developed when using and making were still very close, one that favors on the spot design decisions, needs little or no manufacturing, and therefore implies a mode of control that has little use for today's professionals.

BORROWING FORMAL TECHNOLOGY

Informal products are not easily made by formal procedures and informal ways of working do not seem to inspire formal technology.

Products of formal technology, on the other hand, are used to a large extent in informal making. Indeed, ways of making that were developed in formal technology are often adopted in informal processes.

A case in point is reinforced concrete construction. In the early fifties, when I was a student, this material had just begun to penetrate the education of architects, about a generation after pioneering individuals had used it in defiance of academic architectural standards. We were given demonstrations of the mixing of cement, sand, and gravel and were explained how steel rods could be bent to follow the flow of stresses in a beam or column. All this had the flavor of an initiation into professional secrets. Understanding it thoroughly, we were sure, demanded specialized engineering knowledge.

Most modern products come to us in ways only the specialist knows. But that does not mean that they cannot be used by laymen. Today, in the extensive informal urban settlements of Egypt, where people build for themselves in defiance of urban planning and bureaucratic control, the vault is unknown and mud-brick is of no use for a multi-story structure, but concrete floors are quite normal and concrete beams and columns are frequently seen. These structural elements are usually produced by rule of thumb, without calculation. The work is mostly done by small contractors, if not by users and their neighbors, and there is a good deal of on the job learning going on. Within a few decades, what once was advanced professional knowledge of concrete technology has filtered down to become common knowhow in developing countries.

A worker makes extra money in his free hours to construct a floor in his neighborhood. Eventually he may decide to become a small 'informal' contractor himself. An engineering student who is related to a client may be asked for some help in a special case. All this happens in the social fabric of a neighborhood where people know each other; know they have to live with one another while they try to build, slowly, over many years and from their savings, the solid structure that must secure their children's future. Although called 'informal' the process is under firm social control. Living and working in a same local context makes it difficult to cheat or cut corners and it is hard to walk away from one's mistakes. [6]

Residential construction in most parts of the world takes place in similar processes where no clear line can be drawn between professional and non-professional operations. There is a wide spectrum of alternative procedures between the job that is done with official approval on the one hand, and the job that is purely self-help on the other. It is an area of varied and intricate relations that has never been studied adequately; partly for lack of interest among professionals but also because of understandable reluctance to cooperate by those who act beyond the reaches of the official system. Nevertheless, the latter are in the majority and produce the larger part of the dwelling volume in the world. [7]

This vast, uncharted and unmanaged but ordered market uses products of modern technology without hesitation whenever they are available. Cement and steel rods for reinforced concrete construction are prime examples: they are produced in large plants filled with sophisticated machinery that needed high capital investment. Much more than these two examples is used in the informal sector. There are electric cables and outlets, and plastic conduits for water and drainage. Plate glass and light bulbs are commonly used and soon there will be factory made doors to replace those made by the local carpenter. One can see in villages in the Far East, the Middle East, Latin America and Africa, how in the local smithy steel profiles are welded into shop fronts, windows, and doors when wood is hard to come by. Carpentering a doorframe in wood is probably more difficult than welding a steel one. But the steel profiles are the products of heavy industry. The welding equipment and other electric hand tools like saws and drills are also the products of advanced manufacturing. The tools that first served formal craftsmanship were soon adopted by home owners in the Western countries to fix things on weekends and are now also used in the world's informal building sector.

There is no contradiction between the use of sophisticated products of specialized manufacturing, supported by painstaking design and planning, under the skilled management of workers using specialized machinery, and protected by a world-wide network of patents on the one hand and, on the other hand, the local, improvised, piece-by-piece, not quite legal transformation of a stretch of farmland into an urban environment. There certainly is no conflict of interest between formal factory production and informal on-site building.

But the professional world has yet to learn that it can inherit only half the earth. The lesson seems to get across, although slowly. The board of directors of Tolteca Company, the largest producer of cement in Mexico, was quick to adjust its thinking when the facts became clear during an economic recession. Statistics showed that building construction was down almost fifty percent compared to the previous year. However, sales figures gave only a slump of about twenty five percent in the sale of bags of cement. The discrepancy revealed the informal sector which was not included in the formal statistics.

No one in the company had known how large a part of its production went to customers who build their own homes and buy only one or two fifty-pound bags of cement at a time. The company decided to move on its newly gained knowledge and today (1985), when buying a bag of cement one can obtain for free an excellent self-help building manual, complete with illustrations in color. The printing and distributing of the booklet being a less expensive investment than an advertisement campaign by television. [8]

LIVING AND MAKING

Specialized manufacturing of goods and tools feeds local, improvised developments in a natural way. We find two ways of making things in what, for all practical purposes, is a fruitful, mutually beneficial relation.

The professional is inclined to dismiss the other side as slipshod, wasteful, dangerous to health and incapable of producing anything of quality. But this, as a general notion, is as misguided and simplistic as it would be to blame 'big industry' for the world's inequities or its environmental problems. The issue is rooted deeper. We have been educated in a world of science, technology and organization that is overwhelmingly professional. That culture is, if not ignorant, deeply suspicious of the productive power of the lay world and its variety of social patterns.

There is an unspoken professional ideology in which the physical environment of modern man is viewed as the product - indeed the design - of the controlling professional mind. Those very large areas in the world that are not so organized are seen as aberrations and believed to be in need of replacement rather than just help.

When we trace the route followed by many manufactured things like the steel rod, the bag of cement, the plastic pipe, and the glass pane to find their way in the everyday world, we see them move from one realm of production to another. The one they come from is cast in the professional mode, it is a domain where production is at the center and the social pattern of those who work for it is arranged around it. Individuals enter this domain to play roles determined by the work to be done. It is a domain of production intended to serve a market of similarly formalized building companies. Here design is needed, not only because the technical processes are intricate and in need of detailed instruction, but also because the market out there must be interpreted for the product to be decided upon.

The other domain is one where designing, if needed at all, is still under control of use. In Egypt's and Mexico's informal sectors - as in those of all developing countries - design service is neither available, nor much needed. The user decides with the maker on the basis of conventional form, much in the way we found the farmer and the carpenter decide what to do in the anecdote I reported about earlier. Usually, even in the poorer settlement processes, there is a distinction of making and using but making operates under direct control of use. The creation of things is part of everyday life. The social structure dominates making. We find here a domain of production that tends to constitute itself whenever people inhabit space together.

Both domains have productive and creative power; this point has to be made emphatically. Although we may say that the one is oriented around making and the other around using, they both produce things. Both take part in the transformation of form which is the imprint of life.

The domain controlled by manufacturing, which is the professional one, must inevitably make a social order. This is an order of networks of interests, of expertise and information, by nature not bound to one location. But the other, the domain controlled by use - which is the layman's social order, site bound, culturally specific, and molded by social values - must inevitably make form.

Perhaps this is the lesson we must learn: that social order, to be healthy, must make form and that making, to be done at all, needs social order. Where living is the objective, artifacts will come into existence. Where making is the objective, life will discipline itself around it. We need not choose between the two domains; the question of interest here is how the two can best nurture one another.

Both ways of being with form have their positive and negative sides. Quality is not guaranteed in either one. But it is hard to think of a true culture in which the two do not go together. The purely formalized world where the production of things is no longer a means but a goal, and all life is regimented to make production possible, is decidedly Orwellian. In even the most technologically advanced society, use, eventually, must take over to direct and instruct making. Eventually all things made must settle in a specific place for people to live with them. But today we cannot imagine a life worth living without formal making either. Our daily environment is populated by things that come from a long sequence of transformations by professionally controlled processes before they reach us.

For the professional designer it seems important to understand how the two opposites support and define each other. His intervention is most easily attached to the formal side, but the results of his efforts can contribute to the cultivation of those many forms that will inevitably emerge on the other side where production takes place the age-old conventional way.

TYPE

The brief discussion between the carpenter and the farmer that I reported on earlier could be effective because there was a shared image that made the house already present in a social body before it was actually built. When it finally emerged in physical form, it did not surprise anyone in the congregation.

Likewise, the members of the contractor's team, building townhouses with great speed in a well-organized process could act, individually and collectively, with confidence - much like the workers in the Steinway factory - because each was familiar with the product and understood the sequence of its emergence.

This 'sharing of the form' may explain the creation of houses, towns, ships and many other artifacts throughout history with only a minimal need for plans and models, if any. It also explains how these forms, for all we know, were received by the social body that used them as familiar companions in life; not more peculiar at least than one's fellow citizens.

It seems that the concept of 'type' is closely related to this sharing of images. The type is what all team members know when they work, it is what the workers and the users share when they discuss what to do, it is what the social body has in mind when they expect a new form to come into this world. The type, indeed, is the form that is already with us before we begin. What I called 'conventional form' is often with us in the way of a type. Type is never one specifically defined specimen - it is not a design - but the image of a series of possible interpretations; a family of forms with similar properties, members of which can be recognized immediately by those who know the type.

The concept of type, I suggest, is much more than a means for classifications and more than a way to indicate the historic origins of a form. It is a knowledge of a complex form familiar to a group of people by common experience. Types come and go with the societies and cultures they live by. They are, to a large extent, those cultures.

Of course, if we ask ourselves how it could be that buildings, towns, and ships and other complex, elaborate forms came about in history in the same implicit ways that we recognize today in the informal urban settlements of developing countries, the idea of type by itself does not answer the question. But it may point us in the right direction. Type as a body of implicit knowledge conveniently folded into a shared image, makes us begin to comprehend the workings of a production without designing. How a firm knowledge of the uses of type can be gained from here is another matter. How can we deal with the implicit at all? The methodological questions that arise are formidable and I am certainly not prepared to answer them.

At the same time, we may assume that the uses of type need not be restricted to historical cultures or informal processes in developing countries. The formal and the informal, the tacit and the explicit are not that easily separated. Consider processes that we would not hesitate to call extremely formal, such as the development of airplanes, cars, and the large buildings that arise in the cities of contemporary society - processes where formal designing precedes the making of every part, where all is described and planned and charted - and ask how they can hold together at all. The very explicitness of their designs begs the question. Behind all this formalization large numbers of consultants, researchers, designers, and engineers operate in their own way and it is reasonable to suspect that, to make anything work at all, there must likewise be some shared image of the whole already familiar to all who are involved. A social body's creative process with individuals seemingly acting independently is like building a vault which makes stones float in the air, in apparent defiance of gravity although, in reality, it is gravity that stabilizes the whole. Shared images likewise pull towards a common ground and give direction to all individuals in play while each seems to act on his own, be it in a formal or informal mode.

RANGE OF VARIATION

The map of Pompeii, drawn by Overbeck in the nineteenth century, gives an immediate impression of a consistent fabric. (Fig.1.1) It seems clear that all these houses are variations on a theme. We suspect that there must be a Pompeiian house type, but it is hard to pin it down. In the introduction of his chapter on the Pompeiian house Overbeck offers the picture of the 'ideal type' and explains the names and dispositions of the various spaces that make the whole.

It is not easy, however, to find more than a few examples in the entire city that are close to this ideal. It is somewhat like with the human body. We may see perfect proportions for it, but in daily life we find very different variations.

The search for an ideal form does not seem a fruitful way to understand the type as a generative force. The objective is not to produce the ideal in ever closer uniformity. Overbeck's map is the projection of a three-dimensional reality. The city was an environment of public and private realms, defined by a variety of open and closed spaces, light and shade, color and texture. We see doors, gates, walls, floors and stairs that guide the body's movement and often bear marks of that movement. It is an environment of discreet elements, each the result of a discreet act; the placing of stone upon stone, the placing of a joist, the raising of a column, the closing of a wall, the laying of floor tiles, the application of a color. It is the imprint of an environment in which generations were born and lived, an environment that is never seen as a whole as we see it when we look at the map, but is experienced partially, in sequence, dependent on one's movement in it. It must have been this bodily experience in daily life that gave people knowledge of the type; a knowledge that never needed to be made explicit because the form, thus present in the mind, always coincided with the form present in space. [9]

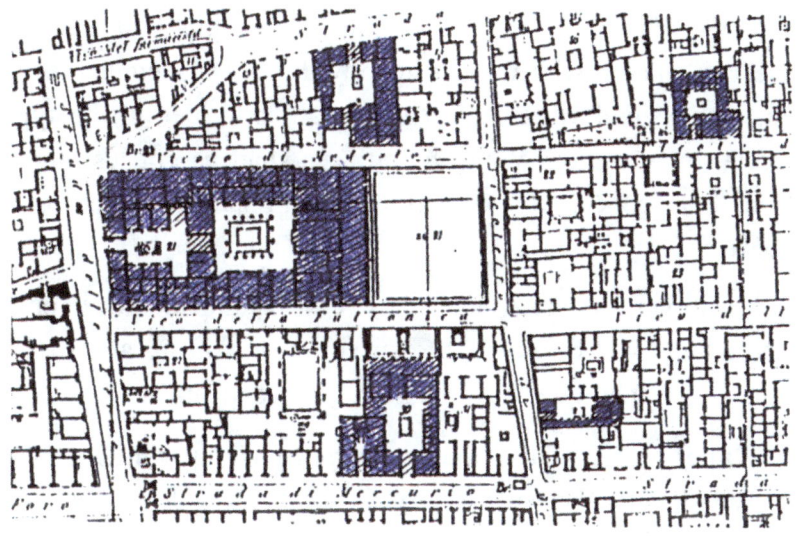

Fig. 1.1, Pompeii, variation in house size of a same type

What first draws our attention in the Pompeii map is the extreme range of the sizes the houses have. Large and small examples coexist happily. But closer scrutiny reveals that the size range has interesting properties. There is a good deal of regularity of room sizes. The large house is not the smaller house blown up, nor is the variation of house sizes a matter of more or less rooms only. If the available land permits, all the different kinds of spaces allowed for in the type are in place. If the house is small, some kinds of spaces are dropped. While the ideal type dictates the combination of atrium and peristyle, many of the smaller houses have only an atrium, but we do not see houses with only a peristyle. In other cases, the latter kind of space appears only partially developed. It seems that the type knows stages of growth. When it expands it spawns new kinds of spaces in specific relation with what is already there. Thus, comparison of large and small variants reveals certain priorities in the type's spatial organization. What the citizens share is not just a static form, but a concept of space arrangement related to the size of the house.

However, the small house is not an incomplete house either. There may well be an ideal to strive for, but all instances of the type are, in their own way, whole, and have identity.

Other artifacts have their own range within which the type presents itself. We have a very different example in the sturdy but graceful flat-bottom sailboats used in the past for transportation and fishing on the shallow waters of the Netherlands and still around today, just for the pleasure of sailing them. (Fig.1.2)

There are distinct types. Each has its own purpose and was developed in a specific region, for a specific kind of water. Like all vernacular forms that proved their mettle, they are triumphs of engineering.

Fig.1.2 Flat bottom fishing boat

These boats were not designed either. According to one witness who gave a detailed account: "The ship was not built from a drawing. It was, as we used to say, 'built from the hand.' The ship builder's eye was all important. He only had a template for the profiles of the stern and the bow, everything else came by itself." And also: "The width of the ship might vary as might the size of the mast. The exact dimensions were often the result of available lumber." 10

One must sail a ship like that to appreciate the precision with which each part of the hull and its rigging is shaped for a purpose; how everything reflects the need to sail that much volume in any kind of weather with only a few hands available.

There is, as in all vernacular forms, an almost palpable sense of the relation between the artifact and the human body. The ship is molded to the body's reach and movement as much as to the forces of wind and water. Being one with the ship, even in its moments of calmness, is what the form is made for. Here as well, within each type very different sizes can be found, but correspondence of an 'ideal' with the real form is much closer than in the house type. The rigging is almost identical for all ships and the differences are in the secondary appliances, real enough in use but making for a less flexible kind of type.

DESCRIBING

The type not only has a range in size and a range of selection of elements that are part of it, there is also variation in the materials used. The flat-bottom ships that are built today for pleasure sailing, although exact specimens of the type, no longer have wooden hulls, but are welded from steel plate. Nylon has replaced sisal rope. Sails used to be cotton; they are now Dacron. We find similar transpositions in building types. The Greek column was a masterpiece in stone cutting, Palladio's Neo- Classical columns are often brick and plaster affairs, requiring a different skill to be honed to perfection. Later, carpenters in North America produced Corinthian and Doric columns in wood and we are once more impressed by the high quality of the craftsmanship.

The point here is not the relative merit of crafts or honesty in the use of materials. Paint covered the veins of the Greek marble columns and a millennium later the plaster columns were painted to look like marble while, still later, the wood was worked smooth as plaster and the same shapes were cast in iron. The tenacity of the image was independent of its materialization.

As we can learn from Palladian architecture, the type can travel. The sharing of images needs no fixed location, only likeminded people. It seems reasonable to expect, however, that presence of a social body in one location for some length of time is a condition for the cultivation of a type. But the fixed location does not seem to be a prerequisite to make it last. People in different locations share customs and religions and also preferences for the forms of artifacts they use. Sharing the form's image in a social body is the primary condition rather than place or period. At all times, however, the form must appear in space as well. To thrive, the type needs to be present in people's imagination as well as in space. Thus, the social body must do both: think the form and make it.

Where it is neither size, nor material, nor a simple fixed shape that defines the type, we find it difficult to describe it. This is particularly true for vernacular house styles. I found that students, when shown examples - for instance, a number of Amsterdam canal houses or Pompeiian courtyard houses or Venetian Gothic palaces - had little trouble to produce another instance of a chosen type, but when asked to collectively produce a description of one, serious differences of opinion arose. A description demands that we agree on what elements belong to the type and what relations among them should be. The students had to reject some features and stress others and finally settle on a way to define what is there. All of that led to long discussions.

What we as observers may agree on, moreover, may not be what the makers of the real-life examples had in mind. Yet for each group of observers, once consensus is reached, the type is real. The type, we may conclude, has no autonomous existence, but lives only in the agreement among people.

Those who made the original objects would have a hard time to describe the type as well, although they would have no difficulty to point out an instance of it if they saw one. One need not be able to describe a type one is familiar with to be able to use it or make it. To seek a description of a type is methodologically in conflict with the nature of the subject. The type is to be acted upon, not to be described. People can know a type very well and share it with others unerringly without ever being able to describe it to their mutual satisfaction. The social body that shares the type need not agree on its description but only on its use in the making of objects. Typology is to steer doing, not description. The type's very implicitness allows it to be complex and rich in details and variations.

To share a type, the experience of living in it or making an instance of it is to be shared first. A description is not needed because it is always present in the forms people live with. We could indeed say that the function of a type lies precisely in that it makes description - and therefore a good deal of designing- unnecessary. When a type is utilized, design thinking can be limited to its interpretation in a specific context; to its adaptation in the real world. The type itself, when applied, embodies all that need not be discussed to begin with: it is the sum of values already shared. [11]

NAMING

Although the type may not easily lend itself for description, it does, however, reveal itself in language. Its parts have names that are known to those who work with them. Of course, each artifact, however humble, must have a name to take part in a culture, but it is characteristic of a type that the names of many parts of it need not have any functional meaning; names can simply denote a form in its context. Ships, for instance, have a 'bow' a 'keel' and a 'stern'. Specific ship types have these parts in their own way, often with their own names. Such names, like the parts of living bodies like 'arm', 'head', 'neck', 'foot', 'stem', or 'root' are, when we think of it, very much carriers of form images. They suggest an order of forms, placed in predetermined relations to one another.

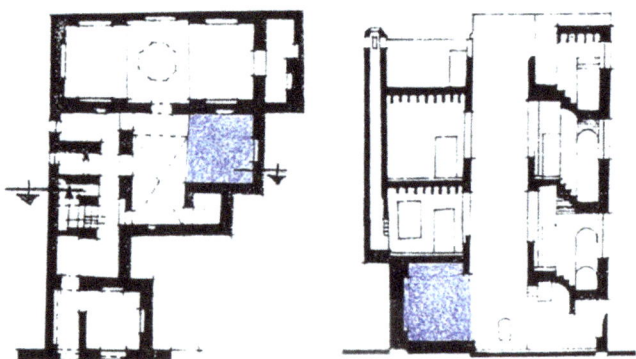

Fig. 1.3, Iwan in an historic Medina hous

Such images of form in a context, indicated by names, are not only of objects, but of spaces as well. In conventional environments we find many of such space names. The 'attic,' the 'porch,' and the 'basement' are well-known examples that convey spaces of a particular kind.

These space names do not convey a function, but an image. We may connect usual functions to a basement or an attic or porch, but it is easy to see that their images would not change if used for something less conventional. In the same way the 'Iwan' in Muslim domestic architecture (fig. 1.3) and the 'atrium' in the Latin tradition are names of spatial forms; they denote form types that are easy to recognize but, in the manner of types, are sometimes difficult to describe correctly.

Typology and vocabulary go hand in hand: each form distinction must be identified. To mention one more example, The Malaysian vernacular house built in bamboo and raised a few feet from the ground has a 'Serambi' which in Western eyes is 'a kind of porch,' but it has truly its own identity in a specific physical and social context. (fig. 1.4)

The names of things and spaces reveal how the forms live in a social body. The vocabularies that go with vernacular forms are therefore extensive and varied and the relations between forms and words are subtle. Shifts in such relations are hard to trace in their evolutions over time because the name is a living thing as is the image it evokes.

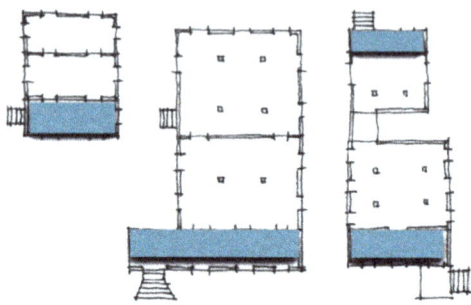

Fig.1. 4m examples of Malaysian Serambi (porch)

The name cuts two ways: it denotes the form and, at the same time, identifies the social body that cultivates the form. It therefore can happen, for instance, that the same form acquires a different name in a given language but in a different region or in a different social group. The name identifies the form and its use identifies a social body that shares its image. In whatever way the relationship between forms and their names may evolve, they always need one another. When the words fade, we can be sure that the forms are soon to disappear as well. Language comes where form and social body are one.

THE LARGER CONTEXT

We found in the type a framework for action and cooperation. Because the type is shared knowledge users can 'think form' with craftsmen and designers. Professionals can coordinate, roles can be identified, and responsibilities can be allocated within the framework of an image that is already alive in people's minds.

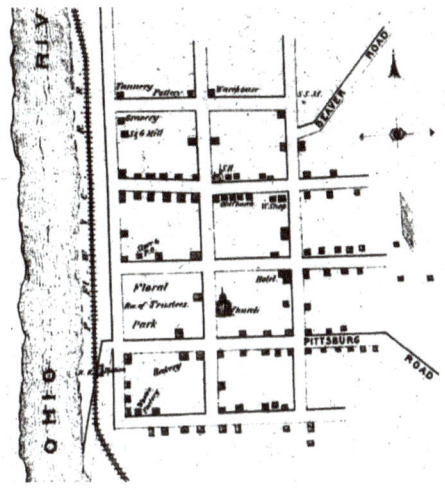

Fig. 1.5 Plan of the town of Economy, 1876

But, in case of a house, how about the larger context in which it is built? John Reps describes the fascinating story of the Oklahoma land rush. Under heavy political pressure the United States Congress decided to make available to settlers the vast planes of Oklahoma although, by mutual contract, these lands had been set aside for Native American tribes. At a fixed date the borders were opened, and people rushed in by train, horse, or carriage to stake their claims for the best parts of the land. Whole towns sprang up overnight. Lots were assigned, streets emerged, and within a few weeks, townscapes could already be recognized in the arrangements of tents, wooden shacks and dirt roads. These settlements, however, did not occupy space haphazardly. They were laid out in a strictly orthogonal pattern of streets with uniform widths and fixed sizes of blocks.

The Oklahoma example is only a dramatic version of a fairly normal occurrence in nineteenth century North America. Towns had to be laid out and the way to do this, as everybody knew, was to make blocks and streets in a grid fashion. The image of the checkerboard pattern was so familiar, and its utility and efficiency were so generally understood that total strangers, converging to the same locality from very different parts of the continent, could stake their individual claims in a wild rush for land and yet conform to a pattern as orderly as if it was all carefully planned beforehand. To be sure, most of the cities of the North American West, as most of those planted earlier on the Eastern Seaboard, were laid out by surveyors and therefore were not exactly spontaneous, improvised settlements. The surveyor's gridded town was a self-evident interpolation of a similar subdivision in a rigidly orthogonal pattern, only to be interrupted by topological features like rivers and mountains. (Fig.1.5) This form is convenient to be laid out without much design effort and it is easy to find one's way in it; both important conditions for a framework in which people from varied backgrounds must be accommodated quickly. [12]

The idea of a preconceived distinction between public and private space, laid out in advance and obeyed in subsequent building acts so that streets and squares make a spatial framework within which individual settlement can take place, is very much part of West European culture. We trace this tradition back to the classical Greeks who were the first to lay out regular street arrangements for their colonial settlements in the Mediterranean. They were followed by the Romans who had their own way of doing it, derived from the military encampment, with its two major axial streets crossing in the center. In the late Middle Ages scores of towns in France were
'planted,' as it was called, by the Church and the Nobility and civic freedom was granted to their inhabitants. Known as 'Bastides' in southern France these new towns, created for economic and political reasons, had rectangular blocks with a town square in the middle surrounded by arcades. From the sixteenth century onwards, the Spanish conquistadors laid out their checkerboard towns in South America. The settlers of the American West were at the end of a period of about two millennia in which a fundamental concept of urban structure persisted, or perhaps was re-invented, as a common idea.

THE SHAPE OF PUBLIC SPACE

The orthogonal grid, as an image of urban structure requires, within the collective imagination, a higher degree of abstraction, it seems, than does the concept of a house or a ship. The ship is perhaps mostly a shape seen from the outside. The free-standing house of the North American suburbs and the farmhouse in the wood-frame cultures of Northern Europe and Japan give us the house likewise seen from the outside, but all imply a spatial order inside as well. A house type in particular conveys an inner composition of spaces that is as familiar to its inhabitants as its outer appearance. Indeed, the Pompeiian house has hardly any outward shape at all. It shares with other courtyard types the introverted nature of this basic form. Here the house is hidden behind blank walls or behind rows of shops, only identified by its gate in a shallow recess. Its image, therefore, is predominantly spatial. It is the form of an environment rather than a shape.

We experience an urban environment somewhat like we experience the inside of a building: finding spaces connected to one another through which we can move. We can be brought, blindfolded, into a familiar town and will recognize it as soon as we can see again, quickly finding our way. The ability to know a form from its inside does not explain, however, how we can know an urban fabric to have a geometric pattern. This requires a certain conceptual distance; we must substitute a more reduced diagrammatic form for the complex reality around us which we never can see as a whole.

In this way the Greek and Spanish colonists and the American settlers shared a more abstract image. Their way of 'thinking an urban fabric' seems fundamentally different from the way Arab citizens knew their traditional urban environment. For the latter, the order appears to be one of spatial hierarchies rather than a predetermined form. Major streets wind their way through the urban tissue to branch to lesser streets that, eventually, lead to dead-end streets and courtyards. There is not necessarily a dominant direction or any fixed geometry. (Fig. 1.6) This is not the place to advance a theory on the Middle Eastern city, but even a cursory comparison with the Western tradition shows that different cultures can adhere to vastly different images of the urban fabrics which they know but never see as a whole.

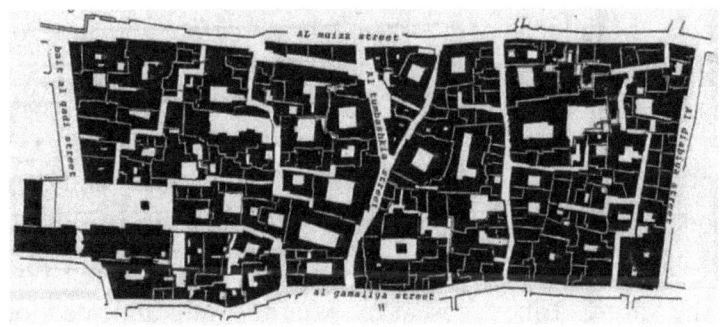

Fig. 1.6, part of old Cairo

The Middle Eastern and the Western urban fabrics differ not only in terms of spatial structure but also by the role of public space. In the Western tradition the configuration of public space presents itself with authority. Public space has fixed boundaries that no citizen is expected to violate. Freedom of transformation on the levels of house form and lot division is strictly limited by the framework of the public spaces. The higher level of public spatial organization is a dominant entity. The party in control of it - the town mayor, the collective of citizens, or the nobility - identifies itself with that higher-level form. We have here the public space as framework, itself the product of convention, within which the inhabitants may distribute themselves freely as long as they obey it.

This autonomous and inviolate public space is in contrast with the Arab tradition of urban settlement. We do not find it in any of the many towns founded after the Muslim conquest of the Middle East, North Africa, or Spain. Nor is it found in what we know of Middle Eastern urban fabrics before the time of the Prophet Mohammad.

It is told how, with the founding of Al-Kufa in seventh century Egypt by the conquering Muslim army, the place of the Friday mosque was determined first, after which arrows were shot in four directions to indicate the limits of the new settlement and the various tribes present in the army were allocated loosely defined areas to settle in.

From what we know it appears that when these settlements grew, the corridors left open between them became major streets. A look at the urban tissues of Muslim towns seems to support this incremental way of finding the larger structure. The approach is no less ordered than the Western tradition of predetermined urban space, but it is the result of a bottom-up development. Patterns of space control determine the shape of the public realm. This is a way of sharing space that is fundamentally different from the Western model where territorial occupation follows the predetermined street pattern. [13]

Comparison between these two ways of going about the same thing needs much more careful attention than it has been given so far. Both systems have proven their efficiency in case of rapid settlement and growth. They are both capable of producing large scale organizations that are understood by their citizens and allow for individual initiative in a larger whole.

The differences relate to fundamental collective images concerning shared space. The Western model assumes a stable geometrically defined form that frames action and control. The form precedes the identification of parties that will build in it. It works from the top down by first setting up a spatial structure to be filled in. The Middle Eastern model, on the other hand, works from the bottom up. It eventually finds the higher level form while individual acts of settlement take place. First the spatial relations of groups of people are determined, then the public structure follows the action. As we will follow more closely in the following chapter, negotiation about space ultimately comes to determine the urban form.

SYSTEMS

The traditional Middle Eastern town is not formless, but its form is fluid and cannot be easily described geometrically. It can be known, however, by its characteristic spaces and the relations between as well as the articulation of their boundaries. There is a hierarchy of streets. One moves from a major artery, through gates into a neighborhood street. From there, passing more gates, into secondary streets that may be dead-end streets or may lead, in turn, to dead-end alleys. The distribution of spatial elements that relate according to rules gives a special sense of coherence, although dimensions and locations are by no means fixed. We gain an understanding of the whole by the recognition of patterns. When we become familiar with such environments, we acquire the image of certain relational principles found in many different configurations but resulting in environments of the same kind. This is the mark of a system: of parts relating to each other by principles of connection and juxtaposition rather than by a predetermined resultant shape.

Here we find evidence of further abstraction of the shared image. The social body, apparently, need not imagine its shared urban form as a shape to be observed from the outside, like seen from the air, to grasp its organization. It can also share the form as a system where the identity of parts and their relations make the whole familiar.

We recognized a similar systemic quality in the Pompeiian house type with its sky-lit atrium and 'Peristylium' surrounded by cells, combined in predetermined ways. The courtyard house type, so much part of the Mediterranean tradition, is a virtual continuation of the urban fabric. It maintains the idea of the more 'public' open space surrounded by more 'private' rooms. In the Roman and Greek tradition, however, we have seen how this basic house form is placed in a predetermined and regular grid of public spaces. In the Arab town we also find the courtyard house but here public space is not only irregular but the result of many individual acts, and much more continuous with the spatial distribution of the house. It is the latter example that makes us understand how people can share the image of a highly complex environment by virtue of its systemic properties. Apparently, the shared image of an urban fabric can be that of a system: meaning that people do share a tacit systemic understanding of the complex environment that surrounds them.

This systemic understanding seems to be something else than the idea of type that we found so helpful in the recognition of house forms, ships, and other composite objects. It could be argued, of course, that all typology is systemic because, when, finally we try to pin down the type, we come to rules about admissible parts and their relation to one another. If we find systemic qualities in all typological forms, we must be careful, however, not to incorporate the idea of type too quickly in a systems theory. The urban structure is, indeed, easily recognized as a homogenous system. It may have a hierarchic structure by itself in its distinction of major and minor streets, but it is basically an image of a single system. The house, however, is not so easily described that way. It takes only a moment's examination to find that the house type offers a combination of systems. Seen as a way of building, the house type yields a technical system; seen as a spatial organization it is another system and its facade reveals yet another one.

In a similar way, ships, planes, and other complex artifacts are complex precisely because they combine a number of distinct, but not always separable, sub-systems. The power of a type is that it unifies this complexity and makes it accessible. We have seen how the type eludes description. It is probably more correct to say that the type, once we recognize it, allows us to see different systems in it; it guides our systemic inquiry. The type, in this way is a vehicle for systemic action but we cannot say that the systems approach 'explains' the type. The idea of system is perhaps more abstract but also more singular than the concept of type. The latter, representing stylistic, spatial and material properties at the same time, suggests a discreet, but intricate and inhomogeneous whole. The system gives us rules for the selection and distribution of certain elements. Following its rules many different shapes can still be made. Both concepts have their place in our collective imagination.

KINDS OF SYSTEMS

Once we recognize the systemic as a property shared by a social body, we begin to see a new layer of bondage of people by physical form.

I have elsewhere argued how the North American 'stick built' construction constitutes an amazingly flexible system shared in the North American society for more than a century.[14]

This principle of building by means of lengths of wood, nominally two by four inches in section, placed at distances of sixteen inches from one another and clad with various materials, is familiar across the continent. It is a truly vernacular system in that it is shared by lay people and professionals. It has been used to build the vast majority of residential volume in the United States and Canada. The 'stick building' system is architecturally neutral. The houses themselves, when finished, can vary greatly in their typology. American Colonial, Ranch-style, Modernist, English cottage, all styles are built in the same system.

This variety of shapes, possible within a same technical system, shows how it represents inner structure. While diverse wholes can be made with it, its power is in its ability to coordinate the process of construction.

The same stick-built system has proven to be a most successful domain for variable materialization. Almost all its parts have been modernized in the course of time. The wooden two-by-fours may be replaced by light steel studs. The wood siding can now be aluminum or plastic. Sheetrock panels have replaced plaster on slats. Insulation of various kinds has been added. The system has proven suitable to receive electrical wiring, pipes and ducts for diverse uses. Window and door frames are easily placed in it; today they may no longer be wood but may contain steel, aluminum or plastic profiles, and so on and so on.

We see here how a system can keep its integrity while transforming almost all of its parts. We already noted the different ways columns in the Neo-Classical system have been materialized. Here along with ship types and house types we find that materials are changeable as long as relations and identities remain. The most durable features of a system are its most abstract ones: the relative position of the parts to one another and the module that governs the placement of its elements

It is difficult to overestimate the power the American 'stick built' vernacular system has over those who associate with it. On the one hand it allows for industrial invention and entrepreneurship. Whomever finds a better way to improve a part in it is assured of a market across the continent. On the other hand, it does not dictate any architectural form. In a way such a physical system, shared by a society, is like a language. It has its limitations of grammar, but one can tell any tale in it. The true language of form, if this too-fashionable metaphor must be used, is technology, not architecture. Architecture is rhetoric; a way of speaking. It is related to the language it uses but remains one step removed from it.

However, many vernacular technical building systems cannot be so easily separated from the materials they use. We recognized this in the account I gave earlier about town houses built so rapidly with so little explicit information. To be sure there was in that case the house type which organized the whole, but while it could have been built in concrete or wood as well in brick it was the traditional brick-masonry-with-timber floor system that allowed its dimensional coordination and enabled the craftsmen to act in concert with such efficiency.

Not all technical systems, moreover, are shared property of those who interact with it. The large concrete panel systems that were promoted in the first three decades after the Second World War in Europe, and which were pressed upon developing countries by manufacturers who lost their European market, never made it into a vernacular acceptance. They were not open to would-be producers of specific parts nor did they offer freedom of invention but, to the contrary, severely limited it. These systems, primitive forms of prefabrication, were neither open, nor industrial and, in their conceptual poverty, gave the idea of system building a bad name. They could not be shared among professionals, nor coordinate professionals with lay people.

For a technical system to become vernacular it must allow some form of control to all parties. The purpose of the systemic must be to simultaneously serve different interests. To the extent that it is successful in doing so it will become a shared property in the minds of people and it will stay alive, growing and changing towards ever richer and ever more effective manifestations.

In the North American 'stick built' system we find the same mutual benefit between formal and informal production as we found in the example of the squatters in developing countries who use reinforced concrete technology. On the one hand there is the world of specialized manufacturing that produces the sheetrock, the metal studs, the aluminum foil, the siding, the insulation, the pipes and wiring, while, on the other hand, there are the home owners, who repair their houses on weekends, discuss changes with a contractor, or indeed build their own houses calling on specialized craftsmen when needed. Truly systemic knowledge is conventional knowledge, as is knowledge of the type. And conventional knowledge is the soil in which our particular designs, professional or homespun, grow best.

THE DESIGNER

Conventional systems, types and other implicit understandings allow us to engage in the making of complex things. They make forms grow and it is this cultivation of form that designing must contribute to in its own particular way. In this light we will examine it further in the following essays.

When we do so we should not forget, obviously, the individual who is designing. But our interest in the designer must be for the sake of the form appearing among us, and not for the individual who engages in the effort to make it enter our world.

It is our habit today to see designing as a personal endeavor; we discuss it in terms of 'meaning' and 'intention' - that is to say we tend to focus on how we feel and what we want when we design. Interesting and important as this may be, it does not help us understand designing itself.

Formulations of meanings and intentions, however brilliantly articulated, stem from the need for self-expression which is a most human urge but often, mistakenly, thought to be the source of art and form.

For a form to take place among people and play a role in society, it must 'come into its own.' As long as it is somebody's expression it remains only a vehicle, a servant, and cannot contribute to the rich and varied culture of people and things. Just as the child must be recognized as a person in his or her own right to become a member of society, so must we allow the form to participate in social space. Hence it is exactly to the extent that a form is no longer beholden to its designer that it can contribute to a culture.

To help form come about we must not smother it with our feelings but try and let it find its place among us. What attracts us in designing is the opposite of self-expression: it is the application of all our knowledge and creative powers towards something outside us. The reward is that, making form this way, we forget our private interests and, to the degree that we help the form and engage in its creation, are liberated from the burdens of self.

Therefore, we do well to discuss designing primarily as proposing a form, an act to be distinguished, but not separated, from our personal thinking, feeling and intentions. This allows us to recognize, finally, how designing makes us join others who will work at, or live with, that same form. What may seem, on first sight, an intolerable reduction of a familiar subject may turn out, after all, to lead us to other, more rewarding views.

two: DESIGNING

THE APPEARANCE

Between the image of what ought to be there and the actual presence of the object which we imagine there is that perilous journey by which we seek to obtain the right form. Our first impulse is to touch the material - clay, stone, wood, or metal - and shape it directly so that under our hands will appear what we already know about but have never seen. In many cases, however, this immediacy is impossible. The distance to go is too long, too many considerations intervene, too many technicalities must be overcome, and we cannot just lay our hands on the world. We sit back to meditate, and, in time, we may first produce another appearance of the form: something halfway towards the thing itself, a representation of both the initial image and the artifact-to-be by means of which each becomes better known. When this happens, we have been designing.

We may make a scale model like those used when the cathedrals were built, or a drawing, or a diagram, or give a written description of the form-to-be. In all such cases the form we seek to make begins to emerge. It is no longer only in our imagination without any presence in the physical world but, at the same time, it is not the actual object that we wish to have either. The form we seek has now appeared in the guise of a representation. This 'appearance' of the form serves as a stepping stone. The art of designing is the art of making those appearances that help us cross the treacherous currents separating the image from the object; the dream from reality.

THE TASK

A design serves a dual purpose. Towards the image - that elusive presence with such power to motivate us, but as yet without true shape - it serves as a checking point. Is this really what we 'had in mind?' Does this really represent what we want? Towards the object that must be made, it serves as an injunction: this is what must be produced.

This dual orientation suggests a primitive model of the design task. In it we already see its most important ingredients.

When we write a sequence of concepts:

Image, > Design, > Object

There is first of all a direction. From left to right we proceed towards materialization. The image, which is immaterial but, I must insist, real because it prods us into action, proceeds, by transformation, towards the right to become a physical object. This, the object, once it is there, will have its own life quite independent from - and often to some degree at variance with - our intentions.

In the opposite direction, from right to left, we could speak of increasing formlessness. We return to initial intentions. It is the direction of evaluation.

Thus we have towards the right:

materialization,

instruction,

_____>

whereas towards the left we have:

abstraction,

evaluation,

<_____

The quality of these directions is the same for the relation between design and image as it is for the relation between design and object. Between image and design we find relations leftward and rightward that are equal to those that we find between design and object.

THE FORM

Let us call the proposal that we make when designing an 'appearance' because it is here that the subject we are concerned with - a house, a car, a machine, a tool - first appears among us. We could say that in all stages of the path towards materialization the form is there. First it is an idea, an image without much shape. The design produces its appearance. Finally, it comes into its own as an artifact.

While in this way the form moves gradually into physical space, it settles in social space as well. To be effective as a proposal, the appearance we produce must be accepted by others; by those who must pay for the product that is proposed, by those who must make it, and by those who may use it. Making appearances under the scrutiny of diverse parties we seek to arrive, finally, at the representation of a form that can be shared by all who are involved.

However, the material object that is produced at last will have its own life. As soon as it comes to share space with us, people may find in it various unintended qualities and unforeseen opportunities. Indeed, the object that is the result of our imagination can, for others, evoke feelings that are quite different from those we carried with us when we were designing it. We have these two extremes: on the one end, the immaterial image which we carry with us before it enters physical space and, on the other end, the artifact that is on its own, no longer beholden to our ideas and intentions. Between these two states of the form we have the design; its intermediate appearance.

In many cases we make different appearances in the same design process. We may use a scale model as well as a drawing and drawings can represent the form in different ways. In fact it is not the media we use that make us distinguish between kinds of appearances but rather the aspects of the form that we may choose to bring to light. A measured drawing for technical specification is a different kind of appearance compared to a drawing illustrating the arrangement of spaces. The spontaneous description by which the client conveys his first thoughts is one; the detailed program that serves as a brief for professional makers is another.

Thus, the string of three may, upon closer examination, be a much longer one, where one appearance follows another like, for instance, from description to program to plan to working drawings. Designing has to do with such strings, their composition, and their use as a path towards the materialization of the form.....or back again to check and reconsider.

CONVENTION

A design is a declaration of intention. As such it can be highly informal; for instance, when we make a quick sketch before we proceed to make the 'real thing.' The sketch is often intended for ourselves, without a wish to communicate with someone else. We seek only clarification of our own thinking towards the form.

Because the form must enter a social space, however, and because its' making usually involves various parties, talking to ourselves soon does not suffice; the products of our designing must inform others. On the right-hand side of the string, closer towards materialization, the appearance must serve to instruct or propose. Towards the left we must check whether instructions and intentions formulated there have been adhered to.

Fig. 2.1, various appearances of the same form

To serve this dual purpose of instruction and evaluation in a professional context we often arrive at a high degree of symbolism in the making of the appearance. The structural engineer for instance, may produce, as a document peculiar to his trade, a diagram in which all parts of a truss are represented as arrows, indicating the direction and magnitude of the forces flowing through the material under certain conditions. The diagram is accompanied by calculations. It is this appearance of the subject in the guise of a configuration of forces that serves as a basis for the specification of material elements - in this case perhaps steel profiles - and their connection to one another: a move towards materialization. The diagram, however, must also be checked against conditions of wind, snow and live loads stated in the program that was written earlier: a move towards the left in the string of appearances.

We see in figure 2.1 three very different appearances of the same form. The middle one shows us the truss as a configuration of forces, the one on the right sees it as a configuration of material elements, and the one to the left shows us the truss as a single element in its environment. On the left side in the string the truss is seen from the outside and, when we move to the right its material parts are revealed.

All three appearances are highly abstracted. To understand them we must be familiar with fairly complex conventions. In fact, to make the necessary calculations we must have studied statics and mathematics and be familiar with the physical properties of the proposed materials. The appearance on the left is based on given climatic conditions and expected uses. The appearance on the right assumes that we know about formally established classes of elements - the steel profiles out of which the truss must be composed - and understand the ways in which they can be connected.

Even this first glance along the lines of a relatively straightforward example makes it clear that the making of an appearance is often a sophisticated affair. It need not always be based on applied science but there is always knowledge shared among the designer and the people she must communicate with - the parties on the left and the right in the string. For this reason, designs are inevitably based on conventions of method and representation.

These conventions must guarantee that the appearance can contain a good deal of information relating to a well-defined aspect of the form at hand, expressed and organized in such a way that it can be absorbed by others with the least risk of distortion and error.

We can see how the act of designing easily brings us among other people. When we engage in it we enter domains of shared knowledge, method, and communication. It takes place in a specific, often precisely defined social, cultural and, above all, operational context which not only determines the way we construct the appearances of the form but also the way in which they are strung together to make a link between image and object.

In all parts of the string, designing appears to be embedded in convention. At the extreme right we may have to do with a multitude of specific details that require the precise coding of elements along the lines of agreed upon modes of visual representation as well as vast amounts of technical know-how. But on the extreme left, where the form invades our consciousness for the first time, we may simply utter a word: 'truss,' 'house,' or 'car' to convey the intention that sets a process in motion. The choice of a word is the choice of a language, a step into a conversation.

CREATIVITY

It seems that the aspects of instruction and evaluation, connected respectively to the rightward and the leftward directions in the string of appearances, are peculiar to designing.

Designers tend to stress, understandably, the creative aspect of their trade and there is no reason to dispute its importance. Those who want to bridge the gap between the image and the object do need their share of creativity and intuition. It takes courage to leap over the abyss of the unknown to get closer to the material form, drawing a sharper profile, adding more substance to the initial image. All the same, creativity is not limited to designing. In all human activities creative people push things ahead, but that does not make them do design.

Artists have been known to make paintings and sculptures without the help of any intermediate appearances of the artwork, not even a sketch. In such a case the person who creates may well step back in concentration to let the image of what has to be made become clear. No single point can be given where such contemplation may first be called designing. To the extent that we think before we act, we work 'by design.' But designing becomes an activity in its own right when we decide to formulate our intentions in a separate document to guide acts by ourselves as well as others. Designing is a special way to be creative. It is directed to the formulation of new information that must instruct action.

As we have seen, this usually happens in a highly formalized manner. Each design profession has its own modes of representation in which specific symbols, words, formulas, colors, and lines are permitted. Engagement in it sometimes occupies our mental faculties to the point that the medium becomes the purpose; the engineer may fall in love with his calculations and the architect with his drawings. We may enjoy calculating stresses in a building structure so much that we welcome the very complexity our design should solve and no longer seek that one, self-evident and elegant solution the creative mind craves for: the form that is only possible in this particular case and which presents itself as inevitable once it is seen.

Technical elaboration in any field offers the opportunity for action within a well-defined, secure context where we do not have to make, ever, the leap into uncertainty that is unavoidable with true creation. Without creativity, therefore, we will make just another journey towards the materialization of a conventional image. But creativity also can lead us astray. It may direct our attention more towards the appearance itself and away from the object which it represents. We may fall in love with an idea so much that its appearance is enough for us. As is the case with utopian architecture, the dream sometimes does not want to materialize. The appearance becomes a creation for its own sake - an object in its own right, unwilling to be a link in a longer chain.

ROLE AND TASK

There used to be a time when, in the design of buildings and bridges, no structural representation, with its calculation of stresses, was made at all.

The emergence of the science of structural analysis, itself triggered by the use of new materials like steel and reinforced concrete, created this new way of seeing built structures. We know how this brought forth a new profession and produced forms mankind had never seen before. The task of building first performed by the master-builder spawned the structural engineer as well as the architect, both designers in their own right. When the structural engineer and the architect cooperate in the design of buildings the two professions contribute to the same string of appearances.

Within the architectural office a variety of design tasks are defined and distributed among different individuals. Specific aspects of a single whole demand appearances that have their own conventions and require distinct talents and skills.

Specialization may follow the distinction of tasks. Site planning, for instance, is now a subject by itself. At the same time, a well-trained architect can do all and, depending on the job at hand, we will still find such a person engaged in the production of a variety of appearances. In other cases, there is a much sharper division of roles within one office.

We can see similar developments towards further distinction of design tasks in other fields as well. In the early days of automobile design little difference was made between the design of the car as a configuration of technical parts and the car as as a device that had to look good and had to accommodate the user. Eventually 'styling' became a skill by itself with its own professionals. Later, ergonomic studies found their place between styling and engineering.

The proliferation of design roles poses by itself one of the most vexing problems in the design of complex objects. When the variety of appearances needed to make a single object increases, demanding different skills and knowledge, their synthesis becomes exceedingly difficult.

With specialization the making of a single appearance may become so elaborate as to pose a full-blown design task itself. The 'logic circuit,' for instance, needed in the design for very large-scale integrated systems (VLSI systems), is only one appearance in a larger process. It is the abstract representation of the routes that signals may follow through the product. In the string of the whole process it is located to the left of the 'layout design' in which the actual spatial distribution of the elements on the silicon wafer is determined. Both the making of the logic circuit and the making of the layout are major design tasks by themselves. They require elaborate and highly specialized efforts, complete with their own strategies and string of appearances. The two together are, moreover, only a part of the much longer string that represents the full design task for an operating chip. Thus, design processes may nest in other design processes. Each stepping stone towards the final goal may, upon closer examination, turn out to be another complete product to be designed. [15]

STRING AND SEQUENCE OF STEPS

To arrive at a complete design, we must know the string to work on. Each design process has its own set of appearances that define the task. There need not be a fixed string for similar design problems. In certain cases we may decide to skip tasks that in other cases are considered essential. Where, for instance, architect and contractor work in close cooperation and mutual trust many details may not be drawn at all. At another time very precisely rendered details of all joints are in order. Due to the close cooperation between client and designer a written program may not be needed, and the shared image may be introduced in a conversation to be directly explored in sketches, plans, and sections. In another case the writing of a program is a separate task for a programming specialist.

Thus, our string of appearances is not a fixed entity by any means. To arrive at a given kind of product certain stepping stones may or may not be used. Each time the appropriate string must be determined. Finding it is part of the development of a design strategy. In case of complex or new problems, deciding what the design task really entails becomes an important part of the total design effort.

In most professional design fields, however, routines for the design task are basically fixed, although they may be implicitly known to a certain extent. When the task and the circumstances are familiar, the string that is needed may indeed be self-evident. In such a case the complex choreography of a design process is embedded in a social body, a situation that is comparable to the examples discussed in the previous essay. And all is well as long as conditions remain stable. Nevertheless, there are reasons to assume that this stability is no longer granted and that we need to know more about designing as a cooperative and dynamic effort.

The appearances to be made can be arranged in the correct relationship of diminishing abstraction and increasing concreteness. The appearance on the left should always be a more general description of the form than the one on the right. The appearance on the right must specify what is determined on the left. The left-hand appearance leaves freedom of interpretation to the one on the right hand. Following this principle we would, for instance, arrange the three appearances in car design mentioned earlier in a sequence giving, from left to right: the system's configuration of major general parts, the ergonomic configuration expressed in distances between parts and their dimensions, and 'styling.' In a similar way we arranged the appearances of the truss in an earlier string.

We can, in general, write the model of the design task as:

I [A1 [A2 [...An [O

in which the symbol [, reading from left to right reads: 'instructs,' and reading from right to left reads: 'is evaluated according to.'

It is not difficult to see how the single string may become one of several branches where, for instance, an appearance A2 instructs A3a, A3b, and A3c. In the same way, such parallel string parts may merge into one. My primary objective, however, is to gain insight into the task of designing and the single string already represents the basic relations to be discussed here. In our scrutiny of the design task we will have several opportunities to come back to the string model and expand it in various ways. It is good to remember, however, that we use the model here as a means towards understanding design relations, but do not want it to be a goal in itself.

The idea of a sequence of appearances, to be sure, does not inform us about the sequence we must follow in the performance of the actual activities indicated in the string. Anyone who was ever involved in a design job of some complexity will see that this model does not give us the work schedule to fulfil our task. The fact that the arrangement from left to right moves from image to object does not imply that our design activity itself should be that linear. It would be quite exceptional if we could run a successful process by first doing the most leftward appearance and next the one to its right and so on until we finally would have, with the last appearance, all that is needed to produce the object, moving all the way without ever looking back.

It is, indeed, very difficult to complete one appearance without working on others at the same time. The engineer making the calculations for the truss must assume the use of certain steel profiles and certain kinds of joints between them to do this. The architect who draws a plan and section for a building, concentrating primarily on its spatial development, must already know in general terms the way of building to be followed. In fact, both the architect and the engineer may, when working on the overall organization of the form at hand, move to the right to go deeper into specific details. Hence the 'principal details' that are so often added to the more global representation of the whole building or the whole truss.

It may happen that insights gained far to the right will lead us to reconsider the initial program placed far to the left. There may be a specific detail concerning material realization that makes us change the very image we started from in the beginning. In general, we can say that the fact that the left-hand appearance 'instructs' the right hand one does not give it precedence in the process of design.

We are used to the notion that command flows from top to bottom, or that decisions go from general to specific, but in design this may not be the best way to go about it. While we work, our attention will shift from left to right and left again. We hop back and forth from appearance to appearance and, as we go, several may emerge simultaneously, with gradually increasing specificity until, finally, all are completed.

How we will move, once the string is determined, to the left or to the right, where we will begin, whether we can approach several appearances at the same time working in parallel, or must follow a sequence, is all a question of design strategy. Our string model does not answer these questions, but it does give us a setting in which to address them. Only the task of what is to be achieved can be given in the model. Against this background the design can proceed.

EXAMPLES

Using the string model as it has been developed so far, we now can consider briefly some simplified cases to familiarize ourselves with the ways we can move, when designing, from one appearance to another.

When we have, for instance, a part of a string of appearances:

$Ax \, [\, Ax+1 \, [\, Ax+2$

which is a section of the complete string that can be given as:

$I \, [\, A1 \, [\, A2 \, [\, ...An \, [\, O$

We can fill in tasks to be done, and take, for instance, a bridge like the one of fig.2.1 as Case 1:

Ax: The overall plan of the bridge in its situation

Ax+1: The static calculations of the bridge

Ax+2: The technical drawings of the bridge

Or we can think of a house as Case 2 and have:

Ax: The house in situation

Ax+1: The house in plans, sections and elevations

Ax+2: The technical details of the house

Let us assume that it is our specific task to make AX+1. It is not important for what follows whether others are working on Ax and Ax+2 or whether we simply limit ourselves at this point to Ax+1. The first thing to do is to find out what information is available from Ax.

In Case 1 there may be a complete design of the bridge's form where all dimensions are given. In that case it may turn out, when we do our calculations, that forces do not distribute evenly. We may conclude that a change in length of some members, changing the proportions of the bridge but not its basic shape, would result in a better distribution of stresses. In that case it must be decided whether changes in Ax will be made.

Still in Case 1, it could be, however, that the appearance of Ax only gives an overall form, leaving open the exact dimensions of the parts, possibly within certain minimum and maximum sizes. In that case the task in Ax+1 would be to search for the optimal dimensions from the point of view of stress distribution. The more general constraints given in this way at the left would leave a wider range of alternatives open to the right.

It may well be that on the right-hand side, in Ax+2, some work has already been done to select desirable steel profiles and ways in which they can be joined. Clearly, some idea as to what materials and parts the bridge should be made out of must exist before the plan of Ax and the calculations of Ax+1 could be made. Since no job is without precedent, bridge design is also done within a pre-conceived technology. There are conventions, based on experience gained in the past, that make us choose certain parts and connect them to one another in certain ways; thus providing a very general and often implicit system. The work in Ax+1 could be done under the assumption that such a system be utilized. When finished, it would provide to Ax+2 a context within which the actual details could be worked out.

We see the same in Case 2. The house in Ax+1 can be designed because there is already a known way to build a house.

It follows that we can work on Ax+2 without knowing anything about the leftward appearances as long as we only figure out ways of building and ways of joining parts. Certain principal details can be made. A range of materials can be selected. In short, a building system may be determined. If that is the case, the work on Ax+1 is the assembly, by means of this system, of a specific form that satisfies the constraints set by Ax.

Systems can be made when no information is available yet from any left-hand appearance, because a system is always a set of rules for the selection and combination of parts: a general way of putting things together. We now can see how systems are a way to pose constraints from the right to the left. In the same way we could, if we wanted to, explore house-forms of a certain type without the benefit of a situation given yet at the left. When working later in Ax we could decide on restrictions of siting - set-back conditions, views, distances from trees and so on - and check by means of the available house typology generated in Ax+1 whether the constraints, thus offered, would be reasonable.

Case 2 gives a good example of a situation in which it is often impossible to complete Ax (in this case the house in the given site) before a lot of work has been done in Ax+1 (the house in plans, sections and elevations). The building as an organization of spaces - walls, floors, roofs and other physical elements - is often intimately connected to its situation.

We see here how the designer may stand between these two appearances and is equally concerned with both. It is a rare case where different individuals would each take responsibility for one of them. Yet, there are two ways of seeing the same thing, and whatever is ultimately decided on the left-hand appearance still leaves a good deal of leeway to the right.

BALANCE

Our model, we must remember, came from the process of immediate making, a process wherein the mind harbors the image and the hands shape the object. The design asserts itself between the two by ever longer strings of appearances. The image - the internal force which drives us to shape things - represents something new, something not yet seen that must find its place in the world. At the same time, however, it must be something familiar. When we want to build a house such as no one ever saw, or produce a car that is truly revolutionary, or a machine that operates on radically new principles, we nevertheless place our image of it in a family of things already known - of houses, cars, or machines.

There is always already the object that we take as a point of departure, perhaps to reject it at the same time. We say: "I want a house, but not...." We say: "This time we will develop a car that...." The image is born from the world as much as it creates the world. From the beginning it is hidden among the things we already know. The vision of something new can only come from what is already known. There seems to be, from the outset, a balance between what is of the mind and what is of the physical world, and the image that makes us bring forth the form germinates from both.

All we can do is to transform what is already there. The temperament of our time is much in favor of the new. Originality is priced high, and it is sometimes believed shameful to admit precedent. This is regrettable because it sets one off in an unbalanced position. Invention connects the new and the old and it is silly to deny the many ways in which all we do and all we are is part of the familiar. True creation is not newness but renewal: a radical transformation of the known. Therefore, the creative mind is irresistibly drawn to observe everyday things. Neither acceptance or rejection are a key to creation. It is the knowledge that eventually all things can change and will change that allows us to participate in the making of meaningful forms.

While it is impossible to deny the world as precedent, it is also impossible not to transform it. The very act of making something is one of making something unique. We are simply incapable of making an exact copy of anything. Only machines seem to be able to do that. Human intervention is inevitably to make a difference, even in spite of itself.

The image and the object give birth to one another. Because we act in time, one step after the other, our representations of things tend to be arranged linearly, but the concept we must embrace is of the simultaneous emergence of beginning and end.

CONSTRAINTS AND FORMS

Let us now examine the situation we find ourselves in when we are working at one particular appearance in the string. Suppose we are working at appearance Ax. Let us say that there is a field DAx which is the domain of our actions. Into that domain information may flow from the outside in the way of constraints. We may also generate new information within DAx itself.

Fig. 2.2, constraints to form relations

Information from the outside, as we have seen, may come from the left (Ax-1) and from the right (Ax+1). The constraints from the left set boundaries to what we can do. They give us the space within which we must make further decisions. The constraints from the right give us, as we have seen, a system - or rather fragments of a system - to work with. We can say that our task is to make with the elements offered from the right something that responds to the instructions given from the left. Let us further assume that the constraints from the outside will not change. The diagram of fig.2.2 suggests that we place all this information in a constraint box C. This box has three drawers. The first drawer - CL - is for information from the left in the string. The second drawer - CR - is for information from the right. The third drawer - CI - is for 'inside' constraints that we will generate ourselves while seeking a solution.

Drawer CI is there because, within the constraints from outside there is always a good deal of room for us to work in. If that was not the case, we could not make a design contribution. We have to make decisions. Some possibilities have to be ruled out; others must be favored. The process by which CI is filled is an iterative one that I will describe in some detail.

Suppose we begin with an empty CI drawer. Thinking about the form we want to make, how it may answer the requirements of CL, and how we may use the system suggested in CR, we may find already some first principles that we want to follow. Perhaps we want to use certain elements rather than others. Perhaps we believe certain configurations of elements will make the form's organization easier. Such ideas are inevitably value judgments that steer the process and narrow the available solution space. To the extent that we can express those preferences in rules or general statements about the desired form, to be followed henceforth, they can be stored in the CI drawer.

Sooner or later, however, we must produce a first form. There is no logical, systematic way to arrive from the constraints to a form. It always requires an intuitive act; a leap into the unknown. But here we go and produce a few sketches that indicate what the solution might be. We will call these suggestions for the appearance Ax that we seek to produce 'Forms.' We will put them in box F.

Once we have some forms, we want to check whether they are indeed compatible with the constraints of box C. Since we do not want to change the outside constraints, let us assume that the forms satisfy those. Our checking now must tell us whether the forms satisfy the constraints in CI.

If they do not, we have two alternatives. We may decide to change our forms. This would bring us into another move towards box F, producing new suggestions. However, we may feel, on the other hand, that the forms violating our own rules are nevertheless steps in the right direction. We will then re-formulate the constraints in CI.

Next, let us suppose that we find our forms in accordance with all constraints that we have so far. The forms in F, as a set, will make us understand better the problem we have to deal with. Looking at them we may find certain aspects we do not like although they are correct within the given constraints. This may lead us to rule out such possibilities and formulate new constraints to be stored in drawer CI.

In a similar fashion the forms in F may lead us to propose changes in the outside constraints when we are convinced that no satisfactory solutions can arise within the present conditions. We may then have to negotiate with parties in control of Ax-1 or

Ax+1. In all cases we see how the production of forms in F may lead to a change in the contents of C.

After such examinations and the resulting adjustments in C we have a new set of constraints to work by. Now we are ready for another leap towards F to replace or modify the forms we had so far. This will lead to another examination and so on, and so on.[16]

EXPLICIT AND IMPLICIT CONSTRAINTS

If we continue this iterative process it will, eventually, yield the final form: the one that we are fully satisfied with. If we would have worked in a strictly systematic way, formulating all constraints that we applied in each step, we would also have, with the final form, a complete set of constraints that would not allow any form other than the one we are now happy with.

The reality, of course, is different. We usually do not make all constraints explicit, nor do we have to. We need only to be explicit when we must agree with other parties on constraints and relations of form. All what is self-evident, all what is convention, and all that we jointly like in the form needs no explicit formulation. The theoretical procedure we have discussed here is nevertheless helpful because it clarifies what basically happens when we design. Let us see what we can learn from it.

There are, in all design processes, two distinct stages that occur alternately. There is the move towards the form and there is the evaluation of the form. The first always entails something like a plunge into the unknown. There is no method that assures us a safe passage from our intentions to a form. The evaluation, on the other hand, allows us to be systematic and analytical.

We will look at these two important modes of action later on. What deserves our attention now are the contents of the constraints box as compared to those of the form box. The forms are the emerging appearance. The final form is, by definition, the appearance that we sought to make. But with that form we also have produced a box full of constraints - rules, restrictions and norms, explicit or implicit - that seem to be a superfluous additional product. On closer examination, however, the explicit constraints are familiar. What we have here is similar to what is usually called a 'program' or a brief: it is like the specifications, rules, standards and requirements that are often compiled at the beginning of a design process to set it on course.

No design process is linear, however. We cannot first complete the program to then have emerge from it, in some mysterious fashion, the form we intended to design. We know that tentative forms, produced while we design, evoke insights that change the program. Today's models in design methodology usually seek to explain this iterative nature of the process. Our cyclic diagram helps us to see how the program is never complete before the design is complete.

Thus, when we look at the finished design and declare that it is 'good' in the way it is, we could say that the truly complete program would be the one that allows this form, and no other. In reality we will, of course, never arrive at a description that is so tight that it will only produce this one solution. There is no need for it, but we know that the first program is only the beginning of a process in which further constraints will emerge together with the form that we search for.

The experienced designer, therefore, tends to move towards the form early in the process. She knows that no accumulation of requirements can drive the design process by itself, but that first sketches, produced early in the game, can produce insight into relevant limitations.

While it is obviously silly to think of a constraint box so full that only one solution is possible, there is also the other extreme in which the constraint box remains empty. This is the fully internalized, intuitive process in which we produce our forms, look at them and try again until we arrive at what is 'good' without ever explicitly formulating any general rule or principle, nor ever trying to set ourselves any explicit limits nor seeking to express values outside of those implicit in the form.

This other extreme of the cyclic process brings us back to the immediate creative act where image and product relate directly through the hands that shape the form. The constraint box comes into play when we need to guide ourselves. Explicit constraints are particularly useful in cases where we do not work alone but must consult with others while we work; a client, a partner, an assistant. In such cases we need to express what values we agree on, and what general principles we want to follow in the next phase of the work. Creation is a lonely act but designing takes place among people.

While the constraint box will be filled gradually - with rules, programs, principles, declarations of intention and agreements - we need to consult with others. But there will be no need to go to the extremes of the model. We do not want so many constraints that no variations of form are possible, nor do we want so much freedom that anything goes. The relation among the participants will decide the use of the constraint box. When there is compatibility among participants, when only a few words suffice to proceed with the making of the form, there is no need to explicitly express many constraints. When, however, the job is more elaborate or when parties hold different values, informal discussion may not be enough and some agreement, perhaps not even written down but expressed explicitly, is needed to proceed. From there, as the complexities of the issues at hand increase, more formal ways of expressing agreement in the way of rules and statements may be needed.

The social relations in the design process will determine the limits of explicit rule making. At a certain point implicit understanding must take over. The forms must speak for themselves at all times and, eventually, all involved must look at them and agree, finally, that this is what should be. At that point the constraints have served their purpose.

SOLUTION SPACE

Suppose we have a simple design problem set by explicit constraints. Say, for example, that a room of fixed dimensions is given. It has a door and a window. In the room a bed, a chair, and a table must be arranged. We now must proceed to a 'solution' called 'bedroom' in which the three objects are given their place. Let us further assume that room and furniture are 'outside' constraints that we cannot change and that we have not yet generated any inside constraints.

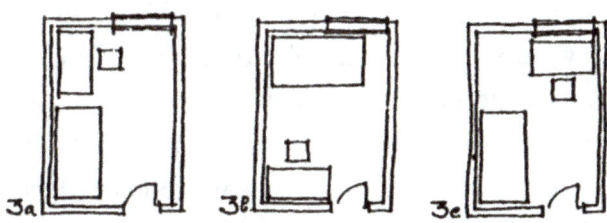

Fig. 2.3 capacity exploration

At first glance this seems to be such a simple problem that we could attempt to make the entire solution space explicit. If all alternative arrangements were known, we could simply pick the one we like best. However, we will soon find that it is not very useful to generate all possible forms without bringing in new constraints. Within the given solution space, for instance, a form in which the bed would be placed against the door, blocking access to the room, would be alright. It is not difficult to think of other valid solutions that we would find equally unacceptable. In short, the solution space is so wide that it allows for a lot of silly possibilities and generating them all, even in this simple case, would produce a large number of useless forms that we would never want to be bothered with in the first place.

It is clearly impractical to try and produce all solutions allowed for by the given explicit constraints. What we really want to do is generate a few of what we believe to be 'reasonable alternatives' to find out what we think is 'really possible.' Suppose we would make three alternative forms as given in figures 2.3. Looking at them we might conclude that we do not want the bed in front of the window, as given in 2.3b, but that the table should preferably be near the window. In this way, observing these first forms made us formulate an additional constraint. We now may produce some new variant while eliminating 2.3b We have here the iterative process between constraints and forms we found earlier. Soon we will have the solution we are happy with.

This simple exercise may make us aware of a few aspects that are pertinent to all design processes

Firstly, there are always, in a design situation, a large number of implicit constraints that are so obvious that we do not bother to formulate them. No one would propose to place a bed in front of a door. This indicates that we always operate in an implicit solution space that is much smaller than the one bounded by explicit constraints.

This really means, once more, that the design process is based on conventions. In a given social body - of designers, of client and designer, of designer and user - many constraints need never be explicit. The stronger the conventional coherence is among those who are involved, the smaller the remaining virtual, implicit solution space will be.

Secondly, the forms that are subsequently generated will reveal, to those involved in the design process, additional constraints that must be made explicit. In our example we could say that we became aware of the desire to have the table at the window only when we saw the alternative in which it was in the back of the room and found it undesirable. Thus, an explicit rule was generated because we were not aware from the beginning that it was a good thing to have the table at the window. Constraints can only come from the consideration of possible - desirable or undesirable - alternatives. Without such alternatives we cannot learn about our values and preferences. To return to the diagram of fig.2: without anything in box F we cannot add anything to C.

Fig 2.4 solution space

We can make a diagram where the solution space available within the explicit constraints (Sa) is given as a bounded area (Fig. 2.4) Within that largest space we have another one, indicated by a dotted line, that bounds the virtual solution space

(Sv) valid for the participants in the design process. Outside Sv lie all those forms that we will never consider because they are implicitly rejected beforehand. When we make forms that propose alternative forms acceptable within Sv we map a small part of Sv which we will call SF.

It is now possible to see how different situations may arise. In theory we will never find forms of SF that are outside Sv because that outside is implicitly rejected by us. However, in practice, it may happen that someone nevertheless proposes a form beyond Sv. That may be the moment when awareness increases, and we may say 'why not?' Or it may just happen that the proposed form is dismissed out of hand as 'impossible' or 'outrageous.' In the latter case some implicit constraints have become explicit.

It may also be that the boundary of Sa crosses the space Sv (Fig 2.4c) This means that constraints given from the outside disallow things that we ourselves would find acceptable. In that case we may have produced some proposals Sf that lie outside Sa. When that happens, we must either accept the given constraints from Sa and reject the forms that violate them or go out and argue for change of Sa. To do the latter we must, as we have seen, enter the larger arena of the design string to influence what happens to the right or the left of where we operate. This may well involve negotiation with another party. It may be, on the other hand, that we are in control of those other appearances as well and can decide to change things ourselves.

Finally, it may be that we look at our forms Sf and decide that some of them are not attractive. We did so when we rejected the alternatives in which the table is not by the window. We decided that Sv should not include these forms anymore. In that case we have found an explicit internal constraint which we may file in drawer CI.[17]

In examples like these we can appreciate the impact of the person who proposes the forms. The designer suggesting the alternative forms illuminates the part of the solution space that we share. His/her choices and insights give direction to the process and must lead us, eventually, to the one form, as yet hidden in the dark space we are exploring, which we will declare to be the right one.

A SIMILAR RELATION

Let us now examine the models we have used so far. To begin with, we saw the design task as a configuration of appearances. This configuration came from the concept of the design as a representation which is placed between object and image. We can trace the path we followed:

First there is the immediate creation where image and object relate directly. It is the interaction between mind and material by way of the hand. We will write this as:

I [O

where I is the image and O the object. In this relationship the design finds its place as an intermediate creation:

I [D [O

From there, we saw how a single representation can only be one way of describing the form we seek to produce. In most design processes we produce several ways of seeing the form. The products thereof are different 'appearances' of the same form. Hence the design process is in fact the production of a series of appearances:

I [A1 [A2 [....An [O

We found that the appearances can be arranged in such a way that the one on the right 'specifies' the one on the left. Hence the left 'instructs' the right. This directional relationship allowed us to consider the movements of the design process as composed of two kinds of acts: the interpretation in more detail of what was seen on the left and the evaluation of what was seen on the right.

Within this general model we looked closer at what happens in the production of a single appearance. We found that here a form had to be produced that conformed to the outside constraints (those found at the left and the right in the string) so that:

$A = C [F$

where F are the forms we generate and on which we work in search of the correct appearance at hand, while

$C = CL$ and CR and CI

in which CL represents the constraints from the appearances to the left of A and the constraints CR are systemic parts, found in the appearances on the right, to be used and combined in the form. The freedom of the designer producing A lies in the relation:

CI [F

where we defined CI as the rules and agreements that the designer of A wants to apply when developing F.

Without any C at all, there is only the relation between the thing we make and the image we work from. This is the 'immediate creation' model from where we started. Therefore, we can see CI, standing between the image I to the left and the forms F to the right, as an expression of our intentions put in the form of general statements, that make a solution space for F. Thus, we can write:

I [CI [F

which is, once more, the design task. Thus, it appears that, within one appearance of the string

I [A1 [A2 [....Wn [O

which itself grew from

I [D [O

another design of the same form is hidden. We therefore can conclude that

C [F

is the most general expression in which all design relations can be summarized. What is on the left always stands for the constraints to instruct the proposals to be generated on the right. What is generated on the right must be evaluated against the constraints on the left. What we called a constraint box is, after all, another appearance of the form, just as the program description is one.

Thus, this general form can stand for the relation between two appearances:

C [F = AX [Ax+1

Or for the leftward extreme of the string where the first appearance begins:

C [F = I [A1

which is very close to the immediate act of making the object without any design as expressed in:

I [O

the difference being that the thing to be made is an appearance and not the object itself.

We can also translate the general expression as follows:

C [F = An [O

which stands for the extreme right of the string, being the point where the production of the object begins.

In all cases, when designing we can speak of the 'leftward' and the 'rightward' directions that we found when we first discussed the string of appearances. In all cases moving towards the left means evaluating what we are doing, that is, to consider the more general framework that instructs our work. Moving to the right means to materialize further, to make the more specific proposal.

THE RIGHTWARD MOVE

The move towards the right always entails the intuitive step into the uncharted solution space to find the form. The outcome depends on the person who makes the move. Given the same circumstances two different people will come up with different alternatives.

Although there is no way by which the rightward move can be made objectively, some degree of method can be introduced. To return to our earlier example, the exploration of the bedroom layout is something one can do systematically. We can begin with a few alternative positions for the bed and follow these with positions for the table and the chair. In this way we could quickly generate a series of forms the differences between which are not trivial. There are other ways to explore a solution space but, eventually, within these methods, someone must generate the alternative forms and suggest the range that is worth an examination. This is a matter of judgment, skill, experience, and talent.

This exploration of alternatives also confirms the fact that the experienced and talented designer always can see alternative forms that make for variations of a theme reflecting her personal preferences within the given constraints.

It is here, it seems to me, that the creative part of designing resides: what alternatives to consider, how to explore the solution space, how to interpret given constraints in a meaningful way. These are the questions that eventually must be answered by a creative spirit. They escape formalization it takes, in addition to experience, training and intelligence, a special gift to see forms in the dark space bounded by constraints.

The intuitive move towards the form is critical in all design processes. To probe its nature, to understand what is going on in the mind that moves into the unknown field of possibilities, to discover laws governing the mind engaged in this adventure, to find what properties of character and sensitivity are best suited for this endeavor; these are all worthy quests of great interest. They are, however, general issues of creativity not peculiar to designing. Although obviously important to designing as well, they cannot be part of the more specific examination we are engaged in here.

THE LEFTWARD MOVE

Once the alternative forms have been generated, their evaluation can take place. It is particularly in this part of the cyclic process that we recognize different kinds of designers. Evaluating the form, we bring to bear ways of seeing that are particular to our discipline. After all, the appearance we are working on demonstrates, itself, a particular way of seeing; it shows the object on its way towards realization.

This is not the place to go into a comparison of the criteria that characterize the various disciplines. It is more generally interesting to identify two opposite ways of dealing with the issue of evaluation, ways that are, to some extent, ingredients of all design processes.

The first is the approach that wants to measure. It is the one that seeks detached observation and comparison. Here we prefer to deal with operations intended to be independent of the individual that does the evaluation. It is the 'scientific' approach where criteria are expressed without ambiguity and can be brought to bear in a formal way.

Of course, that approach has its limitations. At a certain point human judgment must take over. There is not always a formula available to provide us with criteria that can be stated without ambiguity. There are always aspects that cannot be measured but that are very important all the same. Moreover, in most design processes we must take into account such things as the time and the resources available for the job, the client's biases or previous experiences that influence our judgment. One way or another such intangible aspects influence the decisions we must make to proceed. Then there are also those intuitive preferences having to do with the simplicity of the solution or even its elegance. Judgment, in short, is a human activity that cannot be put in formulas or delegated to machines.

Nevertheless, granted that all evaluation is ultimately weighing and judging, there are design processes that rest to a large extent on rational measurement. The performance of a machine, for instance, can usually be quantified, allowing us to establish in an objective way whether criteria have been met. In engineering the circumstances within which measurement takes place can be controlled. Where such conditions are available, we can indeed entertain the notion that the better process of evaluation is the one that is independent of the person who is performing it.

Nevertheless, we know that finally the identity of those involved will make a difference. This is particularly obvious when we consider the design, not of a machine, but of an environmental form such as the bedroom we discussed earlier.

Here we see how the evaluation makes us shift criteria. The design of environmental forms is, to a large extent, the search for criteria. Who shall say that a rule saying that the table must be by the window is correct? Perhaps the user does not want daylight when using the table. Perhaps the view from the window is not that attractive. In a design process like this a criterion is good when those involved in the design process agree it is good and they may change their mind any time. Other observers may come with other criteria; they may indeed 'see' other things in the same forms. Furthermore, what does it mean to have a table at the window? Could the table be placed against the wall adjacent to the window and still be in a correct position? In that case, how close is 'at the window'?

In environmental design the process rests predominantly with the actors who are involved. To be sure, there are objective facts to be recorded: the position of the table in the room, for instance, can be stated in exact dimensions relative to the walls of the room. But such determinations of fact are only a small part of the evaluation process compared to the effort needed to deal with the criteria themselves. To make environmental form is often a socio-political affair. It is by nature a process of agreement and consensus.

The designer of environmental form, therefore, will, in contrast to the designer of the machine, seek to adjust her methods to the human factor. Her ideal process makes herself, client, user, and builder work in harmony to seek the form. Trying to eliminate the individuality of those involved is counter- productive.

Obviously, no design process can be free of either approach. We do well to seek, at all times, ways to express criteria without ambiguity, and to apply them rationally and objectively as possible. Whenever scientific theory and statistical evidence are useful, they should be applied. Whenever objective measurement is applicable it should be used. But at the same time we know that human values have their place relative to design acts.

Judgment cannot be objectified; all designing happens in a social context; often the most important decisions depend on local and temporal conditions; constraints and criteria are seldom cast in stone. There are simply two sides to this coin and no degree of 'scientific' ideology or artistic bias can alter that fact. For some design processes objective measurement and recording of performance are crucial and will demand most attention of the designer. Other processes are dominated by the orchestration of a social context. At some point, however, the other side is encountered, and a balance must be found.

When we recognize these two ways of working, the one seeking to eliminate the human factor and the other seeking to benefit from it, we find that they both have their methodical and systematic principles. Their opposition is not a degree of orderliness. Objective analysis needs a method that is different from the one seeking to arrive at a consensus among actors confronted with alternative forms, developing criteria that may shift later. In both cases, sensible procedures, established with clarity and reasonably controlled are needed. In both cases appropriate methods pay off. For those who are engaged in it, the attraction of designing lies in such opposites. After all, the design process inserts itself between the image and the object. It links the dream to the physical reality and creation to method. It is objective measurement as well as social agreement. The act of designing may have to do with appearances only, making objects that are never quite themselves but reflections of diverse worlds, but at the same time it is an activity where many dimensions of life intersect.

three: SEEING

THEMES AND SYSTEMS

To observe an object, one must enter the space in which it is located, but to describe it one must enter a social space.

When we observe an artifact, we see an imprint of those who made it. Its form tells us about the habits, knowledge, preferences and values of other people. The observed artifact is more than just a thing. It has a social dimension. When, in addition, we describe what we see we choose a certain medium, frame our impressions in a certain way, decide to stress certain aspects, because of the company we keep and the audience we address. In this way both the object and its description are other people. For the same reasons we reveal ourselves, inevitably, by making things or describing them.

The things we make come in groups. We speak of cars, trusses, houses, integrated circuits and towns. Within each of them, included groups are to be found. All these things connect to people who make them, who use them, and who control them. We see power, income, ethnicity or social class in cars, houses and clothes.

We contribute to the manifestation of such implied groupings and their meanings and distinctions by what we make. We build a house, for instance, that has its own identity. We may even consider it unique, but that house is, at the same time, of a certain kind and it establishes ourselves as part of a certain social group. The machine we make, even when it stands for a series of identical forms as is usual for products of manufacturing, is equally significant in this aspect. It is as if we give, in all things that we newly conceive of, our interpretation of a familiar type. While innovating, we always make, to some extent, a variation on a common theme. To the extent that we do so we establish our place among other people.

By making things, we connect them to aspects that are already in use but that we modify constantly. I suggest that we speak of 'thematic systems' when considering such shared properties. They may be called 'style' or 'type' or 'pattern' but always imply rules of selection and combination of the parts making the form: some parts are admitted, and others are not.[18]
Put in the most general terms, what we share always can be seen as a system. A system consists of certain things, and not others, that can be related to one another in certain ways, and not in other ways. No matter that we do it for practical reasons, for beauty, for fashion, or by technical preference, in all cases we choose certain parts and certain arrangements and reject others. To the extent that such choices are based on shared preferences among people who will do their own with them, make their own variations on the shared theme, we can speak of thematic systems.

Being the product of convention, thematic systems rest on tacit agreement. They are, themselves, artifacts and as such must be distinguished from the systems we agree to find in nature, because nature has its own rules that we cannot change. Whatever systemic principles we find in physics, chemistry, biology or astronomy, they are descriptions of manifestations that we cannot choose. With nature we can change our theories but not what we theorize about. In the design of artifacts, however, we obey rules of our own making. The way we build our houses and our machines, produce appliances and tools, and create artworks, we reconfirm themes of form that will be applied as long as they are shared among members of a social group.

SOCIAL BONDAGE

While Thematic Systems are held together by implicit agreement among people, they also hold people together. If the object that we are making is within a shared theme it connects us to a larger world. To that extent our product is a social affirmation. By the use of a style, a fashion, a method, a technical convention, or a symbolic meaning, we identify ourselves as part of a social group.

By making things, we can connect to people far away in space and time. Palladian villa's, churches and government buildings are found on the American continent and Louis XVI furniture is still fashionable in certain circles in France. Apart from its practical purpose the artifact has the power to express cultural ties and maintain old customs. To be sure, in all making laws of nature play their role, but the way we use the freedom left us by nature makes our product an artifact and gives it social meaning. The thematic system is our way to socialize, not by bodily movement or words, but through things.

The concept of the thematic system allows us to make a distinction between physical manifestations - the artifacts - and the social agreements implied in them. Artifacts within a given thematic system are not copies of one another but variations on a theme. I will call them 'variants' within a thematic system. The agreements that we read in them are what the variants have in common: they are the 'structure' of the system.

In this perspective the facades of Gothic Venetian palaces, for instance, are all variants within the system of that name. If we could describe exactly the rules that make a Gothic Venetian palace facade we would have made explicit the structure of their facade system. Given the tacit nature of thematic systems this is often difficult to do. The variants are visible; they are what we make. The structure is found in the variants; it is all what the variants share. It is by nature implicit and can only be explicit by interpretation, the result of which, in practice, can only be approximate at best. Thematic systems reside in a social group, not in an intellectual agreement.

Fig. 3.1 detail of Venice 1539

In contrast to highly technical systems that are fully defined and formally described, the conventional mode of thematic systems makes them powerful and tenacious as long as people keep making variants. When we would try to describe the system called 'Gothic Venetian Palace facade' we might perhaps succeed to our satisfaction, but another group can be expected to take exception to our interpretation preferring a different version. There is no way to offer proof that one interpretation is better than the other. The thematic system is the product of the social body that made the variants. Its structure is 'in the variants.' The structure we hold to be correct depends on the variants that we admitted to our interpretation.

In other words, the variants are the vehicle for agreement. As long as the variants that are produced are considered valid no explication is necessary. But in times of uncertainty, discussion of thematic rules may arise in an attempt to make sure what it is that we agree on and will act on. A design needs to rest on consensus among those involved in the making of it and often some design agreements must be reached in an explicit manner. But these agreements are rooted in deeper and comprehensive implicit understandings.

Formal agreement not withstanding it is by the making of things that a thematic structure is determined. The connection is immediate. We have an instinctive understanding of how to shape things in such a way that social conformation is assured. We spot an exception easily, even when the violated rule was never before made explicit. In North American suburbia front yards have no fences. Would a fence be put up nevertheless, the neighborhood may well feel insulted and take action. We all know similar examples of implicit understandings that became explicit because of deviating variants. It is the violation that produces the explicit rule, which will be expressed by those who want to preserve the tacit structure. We come to discuss the way things must be when an alternative is presented, and doubt arises. Laws, rules, and regulations are there to the degree that deviations give rise to them. All other conformation is in the things themselves. We establish comprehensive, implicit conventions with the forms that we daily use, but which were never mapped.

We already encountered this precedence of the form over the rule - of the implicit structure over the explicit formulation of it - in our examination of the cyclic nature of the design process where the form must be proposed first before we can extract from it the explicit constraints that must guide further work.

The idea of agreement, implicit or explicit, is at the base of the concept of thematic systems. There is no reason to believe that, in thematic systems, there are rules that are objectively right. There is no immutable structure hidden there waiting to be discovered. The formulation that we may attempt is valid or false in the social space in which we formulate it, just as the variant which we produce is appropriate or not in the social space in which it is made. When a form is accepted it affirms, or perhaps re-defines, a thematic structure. When it is rejected the structure is either affirmed or redefined as well. By such acts, the maker either has access to, or is excluded from, a social space.[19]

FORMS AND WORDS

The implicit nature of thematic systems is familiar to us. We instinctively know how to shape things - what things to acquire, what to use and to wear and how to arrange things - to make sure that we are seen in the right group and connect to others in the right way. Conformation allows us to identify ourselves. It allows us to do our own and leave our personal imprint without breaking the common bond. Adhering to an implicit thematic structure we enter into a silent accord with others. We communicate by making our variant form. There is, normally, little need for either party to talk about the meaning of it.

If we find it difficult to put into words the structure of a system that lives by acts of form-making, the problem is, in the full sense of the word, academic. In practice we may quarrel about the admission of a thematic variant, but there is no need for a description of the implicit structure.

This does not mean that the thematic system is something mysterious. It only means that we have here a realm of human interaction that escapes words. Our culture is so strongly dominated by spoken and written language that we tend to forget that words are only one means to interact. Why should we expect that all other forms of interaction - sound, bodily movement, physical forms - can be represented by words? What reason do we have to believe that non-verbal forms of communication are like languages and therefore allow us to 'translate' the 'meanings' they convey into words in the way one can translate a story from French into English? It is legitimate to say that, with physical forms, we can know exactly what they 'mean' without being able to say it in words; not because we lack the skill to articulate it but because spoken and written language cannot convey this kind of 'meaning.' Language is unique, not because it stands for all human experience but because it adds a dimension of its own to it.

Whereas words are spoken or written, a form, any form - the gesture, the sound, the physical object as well as the word - is acted. The act makes the artifact, the meaning is added by the observer.

Forms remain with us after they have been made. They become companions in our lives. We not only see the artifact as the result of an act but also as an opportunity for us to act. It is there to be used, to be changed, to be added to, to be revered, to be maintained, to be abused, to be controlled. By such acts we meet other people through forms.

We can, of course, discuss the form. We can describe it, give our opinion about it and express our feelings relative to it. But that does not 'translate' the form, nor may it explain the act that brought it forth. For the same act, agreed upon by all, different reasons will be given by different participants. When, for instance, people agree to plant a tree, one party may say this is a good thing to do because the tree will bear fruit, another may intend to sit in its shade, yet another may find it adds quality to the environment. It is possible to agree about planting the tree without ever agreeing about possible reasons, motivations and
'meanings' for that act. The form binds people and defines a

social circle simply by being there. What is said about it only tells us something about the relation between people and the form and about their relations to one another through the form.

Thus, by people's comments we learn about the presence of the form in a social context. Regardless of the varied explanations and motivations that inevitably arise around it, the form will exist as long as people agree to have it. In its peculiar manner, it is the ability of the form to connect people with diverse opinions, quite apart from whatever other functions it must perform or messages it is expected to carry. In this way we live with our forms and through our forms as naturally as we live with one another.

The fact that words are unable to cover all of social interaction and that artifacts can add dimensions to our lives that cannot be reached by words is of great interest. It poses severe methodological difficulties for rational inquiry. These should be addressed, not avoided. What we call implicit is what escapes words, but the thematic system is only hidden in that precise and limited way. It lives by people making forms and these forms are, by themselves, quite explicit. The form is there, taking its place in space, and can be understood by all.

IDENTITY

To the extent that the thematic system brings us in a social space it makes us conform, but such sharing of unspoken values is the basis for identity as well. The variant we make is, within the system, unique. What we share allows us to be part of a social group, but our personal variant makes us known as an individual.

We recognize the tree in a single leaf, but no two leaves in the tree will be exactly the same. The systems that artifacts come from make variants, like a tree makes leaves. No two variants need be the same. We are incapable to do the same thing twice in exactly the same way. Even when we intend to repeat ourselves, we may fail to do so. There is always the imprint of the moment. The things that leave our hands acquire, inevitably, some individuality, even in spite of our intentions.

To make uniform things we must, therefore, rigorously eliminate the human factor. Today, the ubiquitous machine has made us familiar with uniformity and makes us cherish the handmade object as something most precious and rare. However, the machine is only one way to eliminate the spontaneous individuality of the human act. In the palaces of kings and pharaohs and in the buildings of great religions we find long rows of columns chiseled out of cold stone by human hands, and all exactly the same. It is for us, inhabitants of the machine age, difficult to experience the awe this uniformity must have inspired when it was produced. Uniformity is something that nature never achieves. Thus, we find the paradox that man, by eliminating in the maker of the thing all individuality, by making himself into a machine, can produce in material objects the very aspect that we recognize, unfailingly, as a token of his presence on earth.

The uniform series of objects denies the idea of theme making on which individual freedom rests. It is therefore oppressive. In thematic systems the individuality of the object reflects the individuality of the maker. Give two people the same theme and they will make different objects. In Gothic Venice each palace was a special place. Thematic systems give us the unity of opposites: We see what is shared and similar but, by the same observation, the identity of each object is established. Each variant comes into its own. We know the system by its constraints but once its thematic structure is accepted only individuality remains.

BACK TO THE MODEL

A more complex object, like a building, is usually more than just a variant in a single system. We can see it, for instance, as a variant of a system of spaces, or see its physical structure as a variant in a way of building or see its facade as a variant of a stylistic system. An appearance in the design process is often specifically focused on a single system in the object we are designing. In that sense we can say that various thematic systems 'reside in' a single object. In fact an appearance is always a representation of parts from the object we are designing, in which we specify their selection and their relations to clarify an aspect of the larger whole. We now can look at the models discussed in Chapter 2 - 'Designing' again, to find how the systems view fits in it.

We found that the expression:

C [F

in which C stands for all the constraints valid in a certain case of making, and F stands for forms possible within those constraints, is the general model of all making from which other, more specific models can be derived. We have the expression

I [A [O

representing the design task wherein I is the image we pursue, O is the object we intend to make, and A the appearance that is the result of our designing.

We found how designing can grow into strings of appearances but that at all times in the arrangement of the string, the relation between parts of it is the same. Thus, the relation I [A is similar to the relation A [O and both are similar to the relation C [F.

This common relation stands for the cyclic process in which we can move, while designing, from left to right and back again. In that process the forms we make can lead to a revision and augmentation of the constraints to the left. The new set of constraints leads to new forms at the right.

This way of 'thinking design' follows patterns that are similar to those we employ when building a system. The constraints to the left lead to the making of forms. To be effective they eventually must tell us which parts are to be used and how they may be related, and that is what a system's structure does. In the forms to the right we will find variants in that system. Thus, we find in the constraints, when properly translated, the structure of a thematic system. The solution space available at any given time in the process encompasses the set of variants possible in the system at play.

Therefore, we can see in the general expression C [F the more specific expression of a system which can be given as:

S [V

where S stands for the structure of the system and V stands for a variant possible within that structure.

However, a system being always an abstraction of the physical reality, a variant V most likely will be a partial interpretation of the form we are actually working on, whereas the structure S probably is only a part of the constraints we are designing with.

We already noted that more than one system can be seen in the final form we are designing, and parts of that form may play in more than one of those systems. The Gothic Venetian palace is recognized by the selection and arrangement of the elements that make its façade, but it is also recognized by its spatial organization which is a variant in a system of spaces. Moreover, there are specific technical features in the way buildings were built in those days that indicate another, technical system. This example stands for almost all making. We usually work with several systems in the same form, each time seeing the whole in a different way.

For these reasons the sum of constraints C in any design activity will, most likely, contain the structures of more than one system.

We already saw earlier that the form F is always more than what the sum of the constraints C want to see; the latter are only aspects of the forms we make, which we use to make sure that the final form answers certain explicit requirements. Our observation of the entire form tells us what we like or dislike in it which, as we have seen, may lead to revision, reduction or augmentation of the constraints. In this way the cyclic process proceeds. We now can see how that same procedure may lead us to a better understanding of the systems we consider in our design.

Being ways of seeing, having to do with values held by people, thematic systems are in the mind as much as in the forms. The fact that in each form a variety of systems can be read does not make them less real, nor the forms more arbitrary. Each system stands for a convention among people and the form that contains the system confirms that bond.

Even when all systems that we possibly could see in a form where identified, we will find that there is more to the form than systems can give us. The systemic approach only allows us to extract aspects of exterior commonality. It is therefore better not to say that the objects that we make are variants or combinations of variants as a mechanistic approach may have it, but to say that we recognize diverse variants from different systems in the objects that we make. There is always something extra in the form, often much more, that escapes systemic description. Systems help us to relate the form to other forms done by other parties or wished for by those we work for or work with. Systems help us to share what we do with others and place it in a social context. What we do in addition is all our own.

HOUSE AND MACHINE

Of the many ways in which we can see the same thing two seem to be of fundamental importance to designers because they relate to two distinct attitudes that are in some measure part of all designing.

These two ways of seeing are exemplified by two kinds of things that we are all familiar with: the machine and the house. We all live in houses and use machines. We cannot escape a relation to either one and therefore the two ways of seeing represented by the house and the machine are based on two kinds of experience everyone has.

I have argued that the first prerequisite to observe an object at all is that we must enter the space that the object is in. Unless we do so we cannot see it, although we may see an image of it, produced on paper, on a screen, in a hologram or as a material model.

I now propose to take this one step further and argue that we always think of objects in terms of our relative position in that shared space. It is characteristic of the machine - the device, the tool - that we look at it from the outside: it is an object to be manipulated. The house, on the other hand, we can enter into. We think of it as something to be inside of. To make use of a house we must enter it. To make use of a machine we position ourselves outside of it.

At this point it is fair to ask: "How about the car, the ship, the airplane? Are they machines that we are inside of?" Indeed, we do make things that have aspects of both machine and house.

Whether we call them habitable machines or moving houses depends on our point of view. A thing sometimes can be seen in two ways and it is precisely different ways of seeing things that we are interested in here rather than a taxonomy of things themselves.

The engineer, whose job it is to design and make machines, will tell us that the purpose of it is to "perform a certain function." The concept of function as used by the engineer rests on an unmistakable and well-defined spatial image. He makes a difference between the thing and it's 'environment.' The machine's function lies in its relation to that environment. The 'performance' of the machine is to be found precisely in its 'behavior' in that environment. Therefore, the concept of 'interface' is important in engineering. It identifies the way the machine is interacting with its environment. It affirms the boundary of the object.

To judge a machine's performance, we do not have to look inside it, but must remain detached observers who see only it's outside. We are part of the machine's environment and prod and probe it to find out what it can do. Most of us use machines we have never seen the inside of. But we 'know' them very well - much in the way we 'know' people - in that we feel we can predict their behavior. The machine's existence is justified by its performance toward the outside: which is where we are!

The technician understands the inside. Although I know my watch as a daily companion, it takes a specialist to take it apart and manipulate its inner configuration. This specialist, to be sure, will ultimately judge everything that goes on inside the watch along criteria of outside performance like we do. For him as well it is the machine's behavior in its environment that it is all about.

In design, the situation is similar. We have designed a machine successfully when it behaves in its designated environment in the way we intended.[20]

As a machine that must meet outside demands the truss that must carry a load is similar to the watch that must tell the time. But the house is a different matter. Here we find it difficult to determine its quality and usefulness by means of performance towards an environment. We do not quite see the house that way. Its purpose seems to be otherwise. Even when we describe the house in purely technical terms - a form that must keep wind and water outside and must insulate the inside from excessive heat and cold, must let in light and air and so on - we find ourselves with an object that must perform towards its own inside.

The house, as all environmental forms, can only be judged from the inside. We must go in to use it, and this prerequisite makes for a fundamentally different perspective. When we seek to apply the engineer's vocabulary, we find that terms like 'interface,' 'behavior,' and 'performance' lose their power. More importantly perhaps, we suddenly cannot observe the whole thing anymore. Once inside we only experience parts of it. Our position towards the thing has been turned around. We are no longer part of the environment of the thing, as was the case with the machine, but the thing has become our environment.

CAPACITY AND FUNCTION

The way we conceive of a thing and make it come about depends on our spatial relation to it. The two ways by which we can relate to the things we make - the outside and the inside perspective - are important because they require different methods of designing.

To understand the difference, we must ask the central design question: how do we find out if what we have designed is 'good?' In the making and designing of machines two aspects govern our evaluation. We look for objective criteria of performance and we expect that we can find, by means of measurement, whether our product can meet these criteria. In other words, the performance of the artifact can be measured against standards that, once they have been formulated, can be applied by anyone who understands the language used to express them. The engineer-designer wants to measure objectively and therefore seeks to eliminate the measurer as a factor in the process. Ideally, once the criteria of performance have been stated, the evaluation should be possible by -- a machine!

With the environmental form this is different. No one will seriously propose that a house is no more than a box with air in it that must offer a desired range of temperature, light, humidity and air circulation. The house is also a configuration of spaces that, by their shape and proportions, their orientation, their materials, the way light reaches them and above all their relations to one another must accommodate a range of activities that cannot be predicted accurately, may shift in time and can differ with each occupant.

An attempt to speak of the object that is an environment in terms of performance and measurement does not lead us very far. Our arguments become elaborate, we must strain our metaphors and the more we proceed the more we have that uncomfortable feeling that we cannot touch what is really important about this kind of object.

It is not so much the behavior of the building itself that we must look at when we want to determine its quality, but the behavior of the occupant. And here again we must avoid making the occupant into a machine with fixed patterns of behavior but must accept that the building will accommodate over its lifetime a number of different occupants with different patterns of behavior. We are precisely interested in the building's *capacity* to be an environment for a variety of activities that may take place simultaneously or in sequence.

It is characteristic of environmental forms that they invite different patterns of use. We would not be correct by saying that the building must 'answer different behaviors.' When saying that, we still want the building to perform. It would be better to say that the house, by its presence as an environment, provides space for a range of human activities - performances in fact - that we cannot all name, nor exhaustively describe.

Architects are very much aware of the fact that the form they design offers space for things to happen. The room, the porch, the balcony, the entrance, the stairs, the hallway, the basement, and the attic are a variety of spaces that, each in its own way, create opportunities for human behavior. What is important about a building is precisely that it has the capacity to stimulate patterns of behavior, which is to say 'performance,' by the inhabitant.

It is this 'capacity' rather than the 'function' that must be assessed. In brief: the environmental form is an object that invites performance and behavior of the people who inhabit it, whereas the machine is an object that performs and behaves. To find out about this capacity we cannot measure, nor formulate objective criteria but must adopt an altogether different method that, for want of a better term, I would call 'capacity exploration'.

With this distinction we come to understand better our attitude, as designers, to the artifact we work on. Our position relative to it will depend on the way we choose to see it. The object itself often can be read both ways. Most forms have something of the machine as well as something of the house. We may speak of 'capacity' when we judge a form's quality as a context for what we do, while we may speak of 'function' when we consider that form's qualities as a thing with which we share a context.

EXPLORING CAPACITY

We say: "If we would place the table here, we could have the cabinet over there and with just enough space left to sit here...." By such mental exercises on possible arrangements for inhabitation we decide whether a room will do for our purposes. This approach is natural to lay people who buy or rent space to occupy. The designer will do well to develop a method of working from that example but will want to evaluate the space on more than one scenario. Given a certain space, we will investigate a space's capacity for various acts of inhabitation that might take place in it. What different uses are possible in a given space and in what way are they possible?

There are two ways to find out how a space can accommodate a certain use we have in mind. One way is to describe that use and ask if the space in question can hold it. Another way is to ask how a certain use might play out in the given space.

Let me rephrase the two questions to make their difference clear. Suppose we consider use as a bedroom for a single person. The first way is then is to ask: "Is the given space good enough to hold a program for a one-person bedroom?" The second way is to ask: "If a person wants to sleep in this room, how could that be done?" Note that the first question evaluates the room against pre-set standards while the second does not. In the first case we must determine what a bedroom's 'program' is. Should it contain just a bed and a chair, or must there be a table and a cupboard as well? If the chosen program can be arranged in the given space in at least one acceptable way, we may decide that the space can be a 'good' one-person bedroom. It is easy to see how this approach demands more and more specification: can the bed be placed under the window or is that not acceptable? must the cupboard be placed only against a wall? etc. etc. We find ourselves seeking the specification of a 'good' one-person bedroom. That is what is called a room's function. It is the functionalist approach which demands statements about what is good or not; this way wants the room to 'perform.' The room is evaluated with a predetermined set of programmatic requirements as a yardstick.

The second question does not try to decide beforehand what a 'good' bedroom is. To answer it we just try to design the room's interior so that a single person can sleep in it. We try to make different variants and by doing so will find out what is possible: whether there is room for more than just a bed and a cupboard or not. While we try to make the most attractive arrangement, we do not have a 'program', nor any rules that can be checked. We do not judge the room by pre-stated criteria but try to clarify its capacity. There will be variants among those that we generate that we personally may not prefer but the point is that there may be inhabitants who will, and in this way, we can leave the judgment to inhabitants or clients.

This way of working is of course more personal: we do want to produce variants that we personally find ergonomically and practically alright. We do apply norms, but they are not programmatic. To the contrary we try to evoke as many modes of use as possible.

In both cases it is of course essential that a space's capacity is explored for a variety of uses. The same room may become a study, or a kitchen, or a hobby room. The very concept of capacity demands variety in use.

In both cases we need to be skilled in developing variants that are meaningfully different. Doing that is something we have not usually been taught. On the contrary, students are encouraged to zero in on the one solution that is 'the best'.

This is not the place to discuss design methodology, but it may be clear that method is useful when we are asked to develop a series of variant solutions in an efficient and meaningful manner. There are for each kind of use always a number of objects among which are pieces of furniture that can be selected and arranged to make one variant and for a second variant a selection from the same pool is arranged in a different way. When doing so we try to follow certain standards of relationship between the objects and between the objects and the parts of the space they are arranged in. In addition, we can expand or diminish the number of objects to be arranged in the same space and this will yield new variants.

This already suggests an orderly approach. There are elements that typically belong to the use we have in mind and certain rules or standards - not necessarily explicit but, for a given designer, consistent - about their relations. There is, in other words, a systemic structure and, in the implicit way peculiar to thematic systems, we recognize that structure in the variants that we can see. Each kind of use, therefore, constitutes a theme expressed in object and their relations and the skilled designer will use it to develop the variants needed to illustrate a space's capacity.

DESIGN ATTITUDES

The architect's way of seeing may be different from the engineer's but the two need not compete. Most things we design solicit both perspectives. The house, as a prime example of the environmental form, representing the inside way of seeing, is also a technical product. As such it pays to scrutinize it as an object that must behave in its environment in predetermined ways. In addition, the house is composed of many parts - walls, trusses, vault, roofs and floors - that must perform in a purely technical manner in relation to the environment that acts upon them and within which they must perform. Their design demands functional analysis.

Cars, ships, planes as machines that represent the best of the engineer's skills have, as already observed before, aspects of environmental form as well. Used in relatively short time spans and for fairly limited purposes, these products are, when looked at in terms of environmental capacity, rather one dimensional for good practical reasons.

The two ways of seeing a form seem to complement each other. Good designers, it could be argued, are sensitive to both of them. We find this complementarity confirmed when we return to the expression of the design task:

C [F

We found how 'moving rightward' involves the development of alternative forms obeying the constraints (C) on the left-hand side of the equation, whereas 'moving leftward' involved evaluation of the forms (F) generated at the right-hand side against the constraints at the left hand.

The two directions imply different operations that require distinct skills. These operations bear remarkable resemblance to those that flow from the two ways of seeing we examine here. The move to the right is an exploration to understand what is possible within the given constraints. We have seen how it involves the exploration of a 'solution space' the boundaries of which are not known. It was necessary to develop a series of alternative forms that, if chosen with skill, would tell us more about those boundaries. The result of that operation depends, so we found, very much on the judgment of the designer who explores the solution space. The alternative forms represent his/ her values. If those forms fit into the solution space offered by the constraints, the latter can be declared compatible with those values.

This is similar to the way of working to explore the capacity of an environmental form where the designer tries to find out whether variants from systems that represent the desired uses fit the form. The important difference is that in the general design model we explore constraints of very different kinds, whereas with the environmental form we find physical constraints only. The environmental form is just one of many kinds of constraints we encounter when designing.

Moving leftward in the general model stands for the evaluation of the forms on the right-hand side. Here we want to find out whether they comply with the criteria stated by the constraints. We must measure and check.

This operation seems similar to what we described as 'the engineer's way'. Both the form and the constraints are now available, and the former must be checked against the latter. This is to be done in such a way that the biases of the operator do not influence the outcome. The results are to be 'objective' in that they can, ideally, be done by anyone who confronts the same information.

We also saw in the general model of the design task, how this leftward move includes more than just checking. After the forms have been tested against the constraints, we can still decide what to do with the results. If, for instance, the forms fail to answer certain constraints we may still decide that we like them and change the constraints accordingly. Alternatively, we may judge that the forms, although they honor the constraints, are not satisfactory and decide to formulate new constraints yielding better results.

Such judgment is part of all designing and it shows, once more, how the human dimension is part of the work in both the leftward and rightward direction of the model.

Apparently, the two ways of seeing we surveyed - the 'way of the house' and the 'way of the machine' are rooted in the act of designing itself. It may be that the temperament of our times makes us inclined to the analytic mode of evaluation found in the leftward move, but the rightward move's exploration of capacity is subject to method like any other operation. It does not allow us, however, to eliminate the explorer as individual.

HIERARCHY

When we consider the various ways we can see the multitude of forms among which we spend our days, we cannot escape the question of complexity.

Fig.3.2 hierarchy of street systems

The acts of construction, building, assembly, composition, and arrangement have to do with bringing together groups and subgroups of physical parts. The task of the designer often includes the organization of many objects into larger, coherent wholes that must answer our initial intention: to make a machine, to make shelter, to perform a function or to accommodate inhabitation. We are soon aware of the fact that such groupings of elements constitute some sort of hierarchy. We place furniture in rooms, rooms are found in houses, and houses align along streets. Any examination of the way we deal with complexity must, eventually, come down to a study of the hierarchic organization of forms.

This is not the place to embark on a treatise on hierarchies we can find in man-made complexity. Yet a few observations about form hierarchies can demonstrate how they influence the relations among designers. The hierarchic order that we see in forms determines to a large extent the way we divide the design responsibilities among intervening parties. They also influence the way we structure the sequence of design tasks to be performed. We intuitively know complex forms to be arrangements of items according to levels.

It is said, sometimes, how "rooms make houses and houses make streets." That may be a poetic way to see things but is not an accurate observation. It confuses two kinds of hierarchies that need to be distinguished. Indeed, smaller physical objects - lower level parts - make larger parts that in turn combine into even larger wholes. This reveals an *assembly hierarchy* which is typical for putting together physical objects.

But in environmental forms we see a hierarchy where an assembly of lower level parts will never make a higher level. For instance, no arrangement of furniture will ever make a room or house, and a city block is not made by an arrangement of houses, but houses are arranged in the city block like furniture is arranged in a room. Apparently, there are classes of elements the arrangements of which takes place in spaces made by the arrangement of other elements and these higher-level elements, in turn, are arranged in yet other spaces made by yet other elements. There are *environmental hierarchies* in which the lower level does not constitute a higher level but inhabits a space made by the higher level. We have a hierarchy of autonomous elements, groups of which are arranged in spaces formed by other autonomous elements.[21]

Now, furniture and houses are very different kinds of elements, but we find similarly autonomous levels when we observe elements that are rather similar in form and purpose. Take, for instance, a network of streets. Residential streets can be arranged in a network of their own. But at a certain point we need larger streets the arrangement of which makes space for a neighborhood with its arrangement of residential streets. In the same way, there is a yet higher-level network of even larger streets - parkways or boulevards perhaps, and so on - all the way up to a network of interstate highways. We have here, once more, a hierarchy where an assembly of lower level elements does not make a higher level.

Fig.3.3 tree system hierarchy

The same property can be found in that prime example of hierarchies: the tree. There is a trunk, there are large limbs, there are smaller limbs and twigs and leaves. The levels in the hierarchy are easily recognized but the leaves never make a branch, nor do branches make a limb.

The autonomy of forms making wholes on different levels becomes even more interesting when we consider the way we talk about wholes in everyday life. When, for instance, I speak of my personal room, I do not mean an empty space formed by four walls, a floor and a ceiling. I mean that space with my furniture and other belongings in it. When we speak of the city block where we live, we do not mean only the four streets and the space they define, but also the houses within the space formed by those streets. Thus, we see 'room' and 'block' as wholes of physical elements on two or more levels. It is the same with the branch of the tree; when we speak of it we do not mean the element that makes a stick but that part with all the smaller parts held up by it together.

The order of built environment, therefore, is not a 'part/ whole' assembly where the smaller parts put together make a bigger one. Certainly, chairs and tables are made of smaller parts and so are houses and streets. But environmental order, as opposed to technical order, has arrangements of discreet forms that make context for arrangement of lower level elements. With buildings, that higher-level context is a space within which lower level parts reside, while with the tree the higher-level form does not enclose but holds up lower level elements. The environmental world is ordered in forms operating on their own level while, at the same time, accommodating forms on lower levels.

INTERVENTION

The hierarchies we perceive in the things we make are of particular interest to designers, because they suggest domains of intervention for them. The room is inhabited by arranging furniture in it. As long as we can freely push around the furniture or hang pictures or put paint on walls or lay down carpets on the floor the room space proper, as a higher-level form, is a constant. It is the result of another level of intervention where, for instance, walls have been arranged by someone else. Likewise, the houses we build in a city block can be constructed or taken down without disturbing the network of streets.

Intervention makes us see levels in environmental complexity. We recognize a level by its transformations. The higher level is, relative to the lower one, a constant and immutable form. It offers the lower level a context for action. When it changes itself, however, its transformation can reverberate downward disturbing the lower level arrangement remains a stable context.

In the case of furniture in the room or houses in a city block, we have examples of a relationship of containment; there is always a space offered by the higher-level form in which the lower level can change. In a similar relationship, the large office building allows for the arrangement of interior partitioning by which rooms are made. Likewise, the residential street finds itself as part of a neighborhood network that can be re-arranged without disturbing a larger network of collector streets.

Fig.3.4 two levels of intervention

But other kinds of form hierarchy, offering levels of intervention as well, are based on different relations between levels. We can think of the tree form again, or of a river with its tributaries, or a network of irrigation ditches, or the distribution of an electrical circuit in a building. In these cases as well we see how higher-level configurations of particular elements offer a stable context for lower level action with other elements. We are free to distribute the water conduits in our house, but when the water main in the street is displaced, our network of the house may have to adapt. In forms like this it is not containment but connection, in this case for the sake of the flow of resources, that governs the relation between the levels.

We see in technical frameworks that serve to hold parts of a complex configuration in place a similar hierarchic dependency in very different forms. The steel framework in a building carries floors, facades and ceilings. In the car the chassis or, more recently, the body frame holds the whole together. With the ship it is the hull that serves as container and framework at the same time. The cast iron engine block receives most other parts that make the engine and the old-fashioned radio had a board on which all parts were mounted, an example followed by the motherboard in a computer.

Frameworks, forms of distribution and forms of containment are found wherever complexity must be dealt with. Because the levels are domains of intervention the relatively stable higher-level form allows us to arrange - and often re-arrange - configurations of lower level parts that actually do the work for which the whole is intended. But when the higher-level form is changed the entire lower level configuration may have to adjust in its spatial arrangement.

It is characteristic of many form hierarchies that the higher-level form serves several, often many, lower level forms. The tree trunk carries several limbs that, in turn, have several branches. The residential street network serves many houses as do the water main and the sewer pipes that are buried in it. The steel or reinforced concrete frame supports many different kinds of lower-level configurations. This distributive property is pervasive: in almost all hierarchic form organizations higher-level forms hold, serve, accommodate, connect, or enclose several, often many lower level forms.

Hence, looking upward in the hierarchies we find that we share, with many other forms, the same services, the same space or the same physical configuration. We speak of infrastructures, frameworks, or public facilities to indicate this property of the hierarchic organization. On any level where we may operate we find, when we look downward, many forms relating to ours while, when we look upward, there is a communal form we share with others.

DESIGNING WITH HIERARCHIES

The levels of intervention that we find in complex artifacts give us the boundaries by which we can separate design responsibility.

When we think of an urban street we think of a complex whole: the road, the sidewalks, the trees, the front yards and the houses all belong to it.

Here we already see three levels. The highest level is the street network that includes our street. Next there are the house lots that inhabit the spaces formed by the street network. In each lot a house is built as a lower level entity, variable in the same lot. It is easy to see that each level is a separate domain of intervention and on each level the higher level sets the stage for intervention at the lower one. The levels constitute a distribution of design intervention by themselves. Even when the same party operates on different levels, the distribution of specific design tasks holds, in that the higher level constrains the design on the lower level. We already have seen that on a given level design intervention can be distributed among many actors in a variety of ways. For instance, a single designer may determine the distribution of house lots, but there are examples where house lots are decided upon sequentially by those who want to build there. In both cases, the level distinction holds.

Higher level design demands consideration of what it means for lower level design. In the example of street, lots, and house levels, the design of the street must take into consideration what sizes of lots and kinds of houses it may serve. The lot sizes must accommodate certain types of houses and in turn the street profile and its dimensions and its detailing want to accommodate the houses.

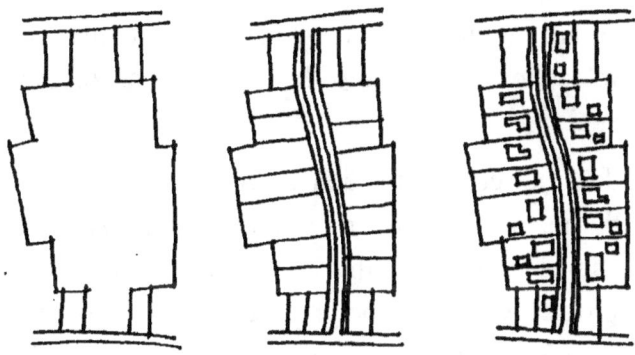

Fig.3.5 *three-level hierarchy: street, lots, houses*

When lots are made narrow, more houses can be built but the character of the street as a whole including the houses will be very different from one with wide lots. An assumption of house lots remains open-ended since house owners may buy several to make larger houses or, conversely, subdivide two lots into three.

This, for instance, is what happened along the seventeenth century canals of Amsterdam. Initial house lots for sale set a standard for what reasonably could be expected to be a 'normal' size house, but in practice the above-mentioned combinations or divisions took place without disturbing the layout of the canals. Nevertheless, our design decisions on dimensions of a street and of the lots along it will be based on such assumptions. This is an example of a capacity considerations as discussed earlier.

Control on one level often can be extended into lower level decisions. Home owners can be legally bound to follow set-back rules to assure a coherent array of front yards. Height restrictions are another common feature to actually assure that lower level action will follow the assumptions of the higher-level design.

This does not change the actual levels themselves but has to do with the identity of parties in control of certain things on a particular level. It may be that the collective of home owners decides on set-back and height restriction rules, and often developers make such rules part of a covenant for the sale of units or lots in a gated community because they know that home owners want to invest in a neighborhood with characteristics they understand and like.

We see how a hierarchy of forms has a certain autonomy that has to do with the properties of physical elements in relation to one another. That autonomous constellation - by definition - allows intervention on each level in the context of the results of higher-level interventions.

Interventions can be made by different parties, but the distribution of control need not be one-on-one with level distinctions. A higher-level party may extend its control in the way of rules and restrictions onto lower level action if a social convention agrees with that. By the same token, higher level control can be vested in the collective of lower level actors who, in that case, formally constitute a controlling party that takes care of the higher-level form that they individually inhabit.

Yet, in this game of shifting control patterns the actual physical hierarchy remains.

CONSTRAINTS FROM ABOVE

The few examples discussed so far may illustrate how familiar we are, when designing, with the hierarchic properties of environmental form. We move up and down the levels at ease. Indeed, hierarchic order seems to compensate for our severely limited ability to comprehend many things at the same time. We make up for it by scanning, like a search light, up and down the levels of form trying to understand a complex whole in its full depth of detail.

The Argentine writer Jorge Luiz Borges wrote about the mysterious Aleph, a secret spot from where one can suddenly see all things in space and time in one gaze. Time and hierarchy dissolve for whoever finds this place and all things everywhere in the world that ever where are seen simultaneously with equal clarity. Unlike the protagonist in Borge's book, we are not allowed to ever find the Aleph. We must forever run the scales of hierarchy (and go back and forth in time) to observe the world we are part of. We only see a fragment at a time and the mind assembles, from those restricted images, a larger one; building up that fuller and richer vision which is never complete but on which we must act.

The observation that levels of intervention coincide with levels of form hierarchy raises the question as to how such interventions relate to one another. Let us examine this situation in the examples of environmental form we used before.

We may identify four levels:

4. The major roads

3. The streets

2. The lots abutting the streets

1. The houses in the lots

This constellation suggests four design tasks. A designer may be responsible for several of these, operating on more than one level in the same project. But whatever the reach of one's design involvement one must deal with each task on its own terms.

Operating on a certain level, we relate upward to the higher level and downward to the lower one. The higher level belongs to the 'site' in which we must work. It is a given and we must explore the options available in the context which it makes. The level lower than ours must settle in the context we provide for it. When we trace the path of the street this implies certain possibilities for allotment. When we decide about a lot size it implies certain possibilities for building in it. Toward the lower-level we offer space and toward the higher-level we explore the given space.

Both directions of action relate to the concept of 'capacity' discussed earlier. We explore the higher-level form to find out what variety of action it allows us. We explore the capacity of the spaces that we produce to find out what another party can do on the lower level.

The same two-way exploration takes place on each level of intervention. The 'inside' view that we found to be so peculiar to environmental form turns out to be the view that designers have on any level in the hierarchy of environmental form when they look upward. There is always a space formed by a higher-level configuration the capacity of which must be explored. By the same token there is always the obligation to assess the capacity of what we offer to lower level intervention.

CONSTRAINTS FROM BELOW

Environmental hierarchies, we have seen, are only one kind of form hierarchy. Other designers find themselves in an 'assembly' hierarchy such as we find in the automobile where the chassis holds a configuration of parts while these parts themselves can be assemblies that have their own framework. The engine, for example, has its engine block that holds lower level parts of the engine in place. The axle, supporting wheels and brake and steering parts serves as framework to another whole. Both are connected to the chassis or the body frame.

Here as well, levels are domains of intervention. But working in a framework hierarchy, in contrast to one of environmental forms, the capacity question may not be the most important one to raise. The designer of the engine is particularly interested in holding in place a number of different parts that must function as a whole. Pistons, valves, and spark plugs must operate in precisely orchestrated unison. Their interrelationship is crucial. The engine block is designed to assure this relationship and make the whole assembly behave as it should.

Thus, we can say that the designer of the engine block is not looking first of all upward in the hierarchy, but downward. The shape of what he is working at on his level of operation is the product of his attention to the lower level parts that must interact smoothly.

In the same way the designer who operates on the level of the car as a whole is actually using the chassis to bring all the sub-assemblies together (or, in the newer car technology, he is designing on the level of the body frame). Here again he is essentially looking down to make sure that the subsystems are interrelated in the appropriate manner and that the framework form adequately accommodates the lower level elements that may already be fully specified, usually by someone else.

Framework hierarchies are forms of assembly and on each level the design must allow for a functional whole made out of the lower level parts. To do so we design the element that serves as framework to hold together all these parts.

We now have found two fundamentally different ways for us to relate to the world of things. The designer in an assembly hierarchy must assemble parts. The designer in an environmental hierarchy must inhabit a space and facilitate freedom of arrangement of lower level parts. Both deal with real physical objects that are given in a particular design instance, on a particular level. In the environmental form the higher level is given. In the assembly situation lower level forms tend to be given. Thus, the environmental hierarchy directs our attention upwards while the framework hierarchy directs it downwards. We found the characteristics of these two positions earlier: the one is the inside view and the other the view from the outside.

We now can see that these two different views are not a matter of free-choice. It is not that the environmental designer prefers to look upward while the framework designer likes to look downward. It is the hierarchies that they find themselves in that pull their attention one way or another because the given elements - forming their 'site' - lie in that direction. We could say that the environmental form explains itself from the top down: hence it is a form of inhabitation of given forms, while the framework form constitutes itself from the bottom up: hence it is a form of assembly.

Thus it is the kind of hierarchy that we find in the form that determines the direction of our design attention and this direction determines, in turn, the methods we will bring to bear when designing.

LOOKING THE OTHER WAY

It is one thing to find that each form hierarchy pulls our attention in a certain direction because that is where the given elements are that we have to deal with. But it is something else to assume that therefore the design task in that hierarchy is strictly one-directional. At this point it is legitimate to ask ourselves what happens when, nevertheless, we look the other way.

The direction in which the given hierarchy pulls our primary attention is where the specific constraints are given. In the other direction, however, we must generalize to learn from it.

Working in an assembly hierarchy we need to accommodate specified lower level elements. When we design a machine, we must make sure that the configuration of its parts functions smoothly and according to specification. We must also know the situation in which our product must operate, but usually we do not find a specific physical context. The machine may have to function in very different locations, but we need to know the context it will operate in. In the true tradition of engineering, as we have discussed earlier, we describe the engine's 'environment' in general terms and proceed to seek the assembly that offers the desired 'behavior' in that generalized context.

Again, it is in the nature of the assembly hierarchy to assemble parts in a functioning whole. At our level of operation, we accept lower level elements as given and seek to assemble a functioning whole. But we know that this form, in turn, will be an element in a higher-level assembly. Part of its function is to perform in different possible locations. An engine, for instance, may be part of different car designs. Therefore, we can only generalize its environment. Our methods are geared to that situation. We measure 'performance' against a precisely specified, but general, 'environment.'

In other words, working in the assembly mode, we work with the method that assumes a range of possible contexts all sharing certain characteristics. We weed out all peculiarities of a specific location.

Once the higher level has been stated, even in general terms, its capacity must be explored as well. Given the generalized 'environment' as part of the performance specification the designer will explore the 'solution space' offered by the higher-level constraints that are assumed to be present. Once the higher level is described there is capacity exploration in the order of assembly as well, but that description will always be a generalization.

With the environmental form which we seek to inhabit we must interpret the higher level as well. But since inhabitation of that particular form is the objective, we will pay particular attention to what cannot be generalized. The bend in the street that we must deal with, the trees in the lot where we will build are important features precisely because they are unique to the situation at hand. In the environmental hierarchy we explore a real form, not a generalization of a series of possible forms.

As we have seen, when designing a form of inhabitation, we look downward as well. When we sketch possible lower level interpretations of the space we have in mind we generalize. When we explore the capacity of a house-lot we will not sketch just one, unique possible house in it. We will search for a variety of typical house forms that may fit on that site. We are not interested in the precise elements out of which these types may be assembled. As far as the lower level forms go, we are only interested in their general aspects.

THE MODEL EXPANDED

The way we choose to make an appearance in the design process depends on the way we see the form which, in turn, brings with it a method of description and evaluation. Our own position in space relative to the forms we conceive and relative to the context we see them in determines the hierarchic order we consider relevant to our task. If the form is seen in an environmental hierarchy, we position ourselves in a given space on a given level. We explore its potential for inhabitation. When the artifact is seen in an assembly hierarchy, we see ourselves assembling an object in an assumed general context. We stand outside of it and observe its behavior. Although the artifact we make an appearance of is not yet in the world, but only in our imagination, we already share space with it and that spatial relationship determines the way we see what we are designing and the way we go about designing it.

In this way the general model of a string of appearances which we discussed in the second essay now acquires an additional, vertical dimension. Depending on the kind of appearance we choose to make, our form not only finds itself in a horizontal string in which the views towards the left and the right have different meanings; but it also finds itself in a form hierarchy. In that vertical direction, however, we must accept the direction commensurate with the appearance's stance. We either see ourselves, by making the appearance, inhabiting a given form, thus looking upward while designing a deeper level of he hierarchy, or we find ourselves assembling parts into a form, looking downwards providing context for lower level forms.

Thus. in the string,

Ax-1 [Ax [Ax + 1

Ax is found in the vertical constellation,

FH

Ax

FL

in which FL and FH are the forms (possible or given) on the higher and the lower level respectively. It means that each appearance that we work on can have relations in four directions. Two in the string of appearances and two in a form hierarchy.

We also found that the actions that we can take relative to the higher or lower level entail operations similar to those connected to respectively the leftward and rightward view in the string of appearances.

When we work in the mode of inhabitation and look upward, we find a given space which we want to consider in terms of the constraints it imposes on our design, like we do when, looking leftward, we want to understand the constraints imposed by the left-hand appearance.

When we look downwards in an assembly hierarchy, we see parts to be assembled; parts that must meet the purpose of the whole, which is similar to our looking at right hand appearances that seek to meet the constraints we imposed.

In the string of appearances, we found that the form at the left 'instructs' and 'evaluates' the right-hand appearance and the form at the right 'interprets' and 'specifies' the left-hand appearance.

In the environmental hierarchy we find that the higher-level form 'serves' and 'holds' the lower level one and the lower level 'inhabits' the higher one.

In the assembly hierarchy we find that the higher-level form imposes constraints to be met and the lower level must meet the constraints we impose.

While vertical and horizontal relations reveal similarity, the differences remain: In the horizontal directions we relate appearances of the same form while in the vertical directions we relate different forms. The constraints in the horizontal order of the string express our intentions, the vertical order of the form hierarchy gives us physical constraints.

four: CONTROLLING

WHOLES

For the eye and the hand sticks and stones are there in an axiomatic way; they are not decomposable. They may be cut but contain no other parts. However, other things, composed of large numbers of diverse elements, things like houses, ships, cars, bridges, or towns present themselves with similar straightforwardness. We have no trouble to identify them in spite of their often-complex architecture. When we reflect on it, it seems truly amazing that we never fail, in the wealth of information that is presented to our senses and in the abundance of impressions around us, to see them at all as discreet entities and not just as a random groupings of lesser objects. There are wholes in the world that contain certain parts and not others, and that have an inside and an outside the delineation of which we unerringly know how to find. These wholes are so self-evident to us that we give them names and discuss their appearance without giving much thought to the fact of their existence in the order of things. Why these wholes and not others?

For a machine, an airplane or a car we may advance a functionalist argument. As we found earlier, the 'outside' view defines things by their 'behavior' in an environment. From this perspective it could perhaps be argued that the object is recognized by its behavior. But we have seen that this criterion is not generally applicable. It will not explain how we recognize wholes from the inside with the same unfailing flair as we do from the outside. A visitor dropped blindfolded in a building or a town familiar to him will soon know where he is. This recognition cannot be fully explained by the objects he sees – the furniture in the building, certain houses in the town - because these may have changed in the period of his absence. The recognition includes the spatial structure that directs our movement, and which depends on our understanding of a wholeness that cannot be seen in its entirety. This wholeness is not seen in an environment, but is an environment, nor does it display any behavior; instead, it accommodates our behavior.

The presence of wholes in our collective existence becomes even more remarkable when we find that they relate by way of patterns of inclusion. The hierarchies that we examined earlier show us wholes, composed of parts, which are, in turn, wholes in their own right. Thus, the ship as a whole may contain an engine as another whole. The door in the wall is a whole in the larger composition of the house. Why does the door seem to be more of a whole to us than the wall itself? The criterion of what makes a whole is not the more or less systematic assembly of parts. Why is a wall primarily a part of something or something made of other parts while a door seems to be something by itself? If the door can be manipulated like a machine, why then is the townhouse in a row of other townhouses or the unit in an apartment building as much a whole though they cannot be so manipulated?

What we know as an apartment is a whole within a larger form. Its outer shape is an invisible part of a large building. Wholes contain other wholes; we find them inside machines as well as within environmental forms. With the same directness by which we recognize the discreet form as an entity, and not as a random pile of parts, we can distinguish, buried within a larger form, other wholes that are known by name but can not be taken out.

We may come closer to an answer to these questions when we abandon the attempt to look only at the artifact but consider once more, as we have done before, our own presence. We share space with the wholes that we observe. They populate the stage on which we act out our lives. There must be a connection between our knowledge of them and the way we live with them.

It seems to me that a good deal of our recognition of order and distinction in the physical world is to be understood from our need to control things around us. What we tend to see as a whole is the form that, in one way or another, serves as a unit of control.

THINGS AND PEOPLE

The idea that control is the common denominator of all artifacts may sound unfamiliar and perhaps even a bit repulsive to some, but there are good reasons to advance it. The issue of control is unavoidable once we adopt the position I advocated in the previous essay on seeing form; that is to say, once we consider ourselves in one space with the forms we discuss. Allow me to explain.

Taking into account our spatial relation to things, we found that the concept of function makes sense when we find ourselves outside of things, whereas the concept of capacity was more appropriate when examining things we were inside of. It was no longer necessary to stretch the idea of function to explain the utility of all artifacts.

This distinction between an 'inside' and an 'outside' view from which to judge things was gained because we decided to introduce ourselves as factor in the study of forms. What things are to us, so we found, depends on where we are relative to them in space. Thing should no longer be discussed unrelated to people.

We come from an academic culture that instinctively separates people from things. Physical science is the study of things. Social scientists study people. This does not help when we want to understand the role of things because in use people and artifacts come together. Nor does this unfortunate separation help when we study the making of things because making is engagement with the thing made. The same applies to designing.

The activity in which we 'think' forms that must be made and used cannot afford to separate people and things either.

Now that we have entered the same space the form is in, we must take the full measure of that position. Our relationship with the form is not passive. It is intolerable to live only with things that we cannot manipulate. We seek to 'take matter in hand' and like to be able to put our mark on the environment we live in - take things with us, store them, distribute them, determine their place within our lives. We want to be able to transform some part of our environment for it to be more than a prison. Once we have entered the space inhabited by artifacts we must, to inhabit it ourselves, transform to some extent the state of things, and transformation is the result of control.[22]

ATTACHMENTS

The most important roles by which we relate to the artifacts that we surround ourselves with - at least for the purpose of our inquiry here - are those of the user, the designer and the maker. Each role attaches us to things in its own way and each way of control has its peculiar aspects of transformation, contributing to humanities' ceaseless creation and re-creation of things.

Use as a form of control is more direct and meaningful than ownership. It cannot have the latter's abstract qualities but requires a commitment to identify ourselves with the things that we use. Use implies manipulation which, in turn, causes movement and also wear and tear. Indeed, in most cases the act of using involves transformation; the closing and opening of doors, windows, boxes or books; the driving and maintenance of cars, the arrangement of furniture and the application of tools are only some of many examples of the almost inevitable link between use and the change of forms.

With such transformations, the shift from use to making is most naturally made. Use, maintenance, repair and making form a cycle of continuous changes by which we live with artifacts.

The designer's job is to be in control of the form before it is an object in real space. The design, when it appears as a model, a drawing or a program, is a thing by itself and subject to the physical constraints of all making. Relative to the appearance the designer is a maker - he produces an object. But his product is also a proposition for another making, this time not by the designer but by someone who will execute the design.

In a similar vein the designer will not expect to use the artifact the image of which he made appear. Thus, the designer, as far as the form he is designing is concerned, although in control of its emergence for some time, is one step removed from making and using, but serves both.

Although the designer's control of the form may be the more immaterial - but not less real - one, the maker's control is perhaps the most physical and dramatic of the three. Making involves the immediate and forceful transformation of matter. It is the tangible side of creation where things acquire their shape in a deliberate and organized manner. The sheer effort of making is a vital ingredient of life, quite apart from the product's purpose or its merits. To have made something is an experience we do not easily forget.

But the professional maker does not control the thing he makes for very long. Once it leaves his hands the artifact begins its existence under the control of others. It will usually change hands more than once and, over time, will find itself in the orbit of different powers. Our artifacts have their own histories.

The transition of a thing into another sphere of control has its own significance, be it a cool commercial transaction or the pregnant passage of a gift. When such changes occur in our relations with things there may remain, with the object that we part with, something of an attachment if not an identification. We encounter with special feelings the places where we lived in the past or the things that belonged to persons dear to us. We do not only note in their present state the lives of others but there remains with them, only noticeable to us, a scent of the past. Things in their passage through tine and space - houses occupied by successive generations, cars traveling the roads, ships cruising the seas, books inhabiting shelves - have a bond to diverse, often unrelated people.

Something of our recognition of the artifact's individual existence lingers with all things we encounter. Love and care bestowed on things for the sake of their being there, being part of our lives, is universal and independent of their function. It is, for example, the source of decoration: when the work is done, we cannot let go of the thing under our hands, and we embellish. It is in such mute attachments to artifacts that we emphasize how artificial they really are, and at the same time leave our signature, giving testimony of our relation to the object.

Our first attachment to things may well emerge with the act of recognition. To notice something is to take it into the realm of our life. In that way the tree we know from our walks in the woods becomes part of our world. Next, we may arrange natural forms we find in nature. The rock in a Zen garden, in its immutable autonomy - being nobody's servant and nobody's master - is nevertheless an inhabitant of the world of human beings. In the recognition of things, the distinction between artifact and non-artifact fades, and patterns of control may expand accordingly.

It is perhaps our deep, instinctive drive to link control to all recognition of things, instead of motivations like greed and sensuousness, that makes the sage renounce all attachment to them. Withdrawal need not be a refusal of the order and beauty of all existence, nor a denial of care for artifacts, but is a response to the bondage of control. The things we control invade our lives in more than one way. To be able to share space with things, to enter the space things are in and accept their presence while nevertheless remaining free, takes a good deal of wisdom.

THE USERS WORLD

Where use, making and designing are three ways of exercising control over the forms we make, we can expect to see differences in forms that are created when one or the other is dominant. Is it possible to see, in the artifacts that we observe, what party was predominantly in control at its genesis, the maker, the user or the designer?

Fig 4.1. Part of the excavation of the city of Ur, ca. 2000BC

Ours is a professional world and we tend to see artifacts primarily as the product of designers and makers; created with the user in mind to be sure, but firmly under professional control. Use in the functionalist approach is a passive concept. The user is studied, is served, indeed pampered in a competitive market but is mainly considered in an abstract way. He is expected to push buttons and turn dials, but his performance is not expected to shape or transform what he works with or lives in. In the past use was there first. It spawned making and, much later, designing. To get a feeling of what such a use-dominated world is like it is worth observing the tradition of Middle eastern urban form. The urban fabric revealed by Woolley's excavations at Ur (*Fig. 4.1*), showing a city of some four thousand years ago, is strikingly similar to that of old Tunis (*Fig 4.2*) which is, today, typical for a large number of extant urban fabrics within the Muslim tradition.

The configurations of such urban fabrics have long been regarded by Western scholarship as 'unorganized' and even chaotic but, as we might expect from such a long tradition, they are highly ordered organisms, the forms of which are not random at all. But their ordering principle lies with the primacy of use. Basically, Koranic law states that those in control of real property are allowed to transform their houses in any way as long as no harm is done to anyone. What constitutes harm is a matter of convention. For instance, in Muslim culture, visual privacy is extremely important, and one is not allowed to violate it in any way. Hence the opening of a window in a wall, allowing a view into a neighbor's courtyard, will cause harm as will the building of a roof terrace without a wall around it.

For the same reason a new door into the street must not be placed opposite an already existing doorway. However, one may have one's floor beams resting in the wall built by a neighbor. One can also freely build out into the street as long as traffic is not blocked. One can build over an alley to add a room to the second floor of one's house and place supporting columns in front of the opposite neighbor's street wall. One can even block a street and create two dead-end streets by building across it as long as neighbor's and customary passers-by agree.

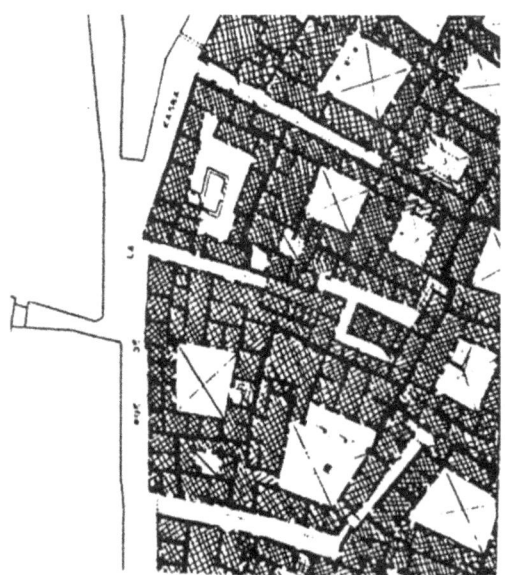

Fig. 4.2 Part of Tunis Medina, dead-end alleys shared by clusters of courtyard houses

The traditional Muslim urban fabric was in continuous piecemeal transformation under the initiative of those who actually occupied and controlled the built forms. There were no predetermined public spaces with inviolate boundaries behind which private forms must remain. Nor are there any general rules that normalize forms. The entire process is based on consensus and individual responsibility.[23]

We may safely assume, given the similarity of traditional Muslim environments to those built much earlier in history, that Koranic laws, pertinent to extant living fabrics, reflect much older customs.

These intricate urban fabrics are the result of principles of procedure rather than principles of form. Neighbors must try to resolve their differences by mutual consultation and agreement. If need be, they should call other neighbors to arbitrate. In each dispute all who consider themselves affected by the disputed action are allowed to participate in the deliberations. The process of continuous transformation of physical form is mirrored by continuous dialogue within ever-changing groups.

The Middle Eastern traditional urban fabric is an example of environmental form produced by control at the lowest possible level of the social hierarchy. The process giving rise to these complex and ever-changing forms is one where use dominates: where all making and all deliberation prior to action is always under immediate control of those who use.

In this user-based procedure there are no general rules of form. The building code as we know it today is foreign to the

Muslim tradition because it must necessarily generalize form. It must say, for instance, that no window may ever look into a neighbor's courtyard and no entrance to a house may be opposite another house door, but in the traditional Muslim fabric it remained possible to have a window overlooking a neighbor's courtyard if neighbors have come to an agreement about it. The window may be too high for the occupant to look through it. It may have been there before the courtyard was made and therefore the user of the courtyard harms himself but is done no harm. For similar reasons doors may be found, occasionally, opposite each other in a street.

It takes a generalizing way of thinking about form to see public space as constant and inviolate as is the case in the Western European tradition. General rules of form cannot regulate the shifting of boundaries between public and private space as will occur when one builds out into the street or across it. There are rules between those seemingly free flowing forms but they deal with the behavior of the people who control the forms, not the forms themselves. These rules tell people how to consult with one another. A neighbor's attempt to build across the street may be stopped by some who find their way to the mosque blocked, while in another case the same act, resulting in a similar form is approved by those who find their path likewise obstructed. Local agreement determines the outcome, the forms are the result.

We must, to regulate form making, first think of use as different and separate from making and designing. The form rule raises the question of use precisely because it separates use from making and designing, and places control in the hands of the latter two. The form rule is technical and professional by nature.

The social rule, the rule of procedure, delegates control. It makes a general statement about where control should reside. The form rule, on the other hand, claims control for the regulator. To the extent that the form is predetermined, control is taken out of the local, ad hoc, situation; yet it is in the local and the ad-hoc that use resides.

THE AUTONOMY OF MAKING

Once we want to make something for others, perhaps unknown to us, the question of use arises. The distinction to be made is not necessarily one between roles but, again, one of control. The patron in old Tunis who decided to build or transform his house would call in carpenters, masons and other craftsmen. Like in all vernacular building the scope of the crafts must have been shared by the users, but the skills to execute them were certainly held by distinct individuals from early times onwards. In vernacular tradition the user controls the maker and therefore, ultimately, the form. Surely, Thomas Jefferson, to give a similar case from another, non-vernacular context, did not build Monticello with his own hands but we say with good reason that he built it. Christian Huygens must have called for the help of the local blacksmith and the toolmaker to experiment with his clocks, but they are his product.[24]

The question of utility becomes the concern of the maker not when he acquires his skills and becomes a professional, but only when he decides to apply such skills on his own initiative and seeks to produce in anticipation of use rather than in direct response to it.

For the maker to take the initiative, he must guess what the user needs. The user becomes a hypothesis rather than an individual. This step towards a theory of use enables manufacturing and modern mass production. It is entirely appropriate to call the age of manufacturing the age of functionalism because it is the concept of function which represents the generalization of use which, in turn, enables making to come into its own. Nevertheless, the predominance of manufacturing today must not make us ignore those forms of production where the question of use need not be posed but where skill, knowledge and control over form are not less advanced.

Inhabitation is one instance where the question of use need not be generalized. The traditional Muslim urban fabric, mentioned above, may be an extreme example but, as I argued in the first essay, it is generally true that settlement forms, being local events, have strong links with user initiative and tend to escape dominance of manufacturing.

There is another instance where the question of use need not be raised, in this case not only because there is no need for generalization but because there is no need for use at all. The work of art, as we define it today, recognized in its own right, for its own sake, is also the result of making that eludes the generalization that manufacturing needs to take initiative. There is no other maker who claims control over his product after it is finished and after it has changed into others' hands, but the artist. All other making assumes manipulation and ultimately transformation of the artifact by subsequent controlling parties. But no art collector today would think of changing the art object he owns. No painter or sculptor today would feel free to transform the work of another artist. If control, as distinct from ownership, resides in the act of transformation, control over the artwork remains with the artist. Because of its total subjection to its first maker the work of art has become the ultimate assertion of making. The art object, as we know it today, is there because it has been made. It offers no other justification. Indeed, it refuses any other obligation, be it social, technical, economical, or legal. Particularly the traditional religious constraints are rejected. It is the act of making, be it personal or collective, that is the ultimate reason for the modern artwork's existence and therefore any transformation by others will destroy it.

This claim for ultimate control by the maker seems to be modern. In history inviolability rested on other grounds. Worship must have been one of the strongest reasons for permanency of the art object. Power and its defiance of death, another. Love and the desire to prolong remembrance yet another. In such cases it is the social context of the artifact, its residence in a public realm that protects it and assures its preservation.

To the extent that there is a social space for the modern artwork it is one that is institutionalized for art itself. The public places for the vast majority of the artworks today are the museum and the gallery. Art, more than ever before, feeds on itself and, by doing so, reinforces the claim of making for making's sake.

BORROWED CONTROL

Where making is immediately responding to use it is the patron who 'thinks form' which is why we feel Jefferson 'built' Monticello and Huygens 'made' his clocks. When making is emancipated, as is the case in manufacturing, it is the maker who bears responsibility for the form. In those cases, the user as well as the maker can be the designer. In both cases as well designing can become a pronounced, well defined activity in which appearances are produced to guide the emergence of the form, but the act of designing need not yet be a separate role played by a separate individual.

It is only at the point where design becomes a professional activity of its own, that control becomes an issue.

We now can see how the sequence from image to object which we discussed in the second essay as represented in

I [D [O

where the design activity D, finding itself between the image and the thing-to-be-made, easily becomes a sequence of control.

At a certain point some party decides to invest in the creation of a form for whatever reason or objective. There must be, at least, a first image of something that wants to appear in the world but has not yet been seen there. The first transition of control over the form takes place before the form has ever been physically manifested: a designer is asked to produce an appearance of the desired object.

The control of the designer is real. When the image of the form passes on to him, he is the only one to transform it. But, as soon as the appearance is there, the designer has no power over it and is not the one to decide what will happen with it. Control is delegated to him for only a limited period.[25]

Not being the maker, the designer must in his representation give information about what is to be made. Not being the user, he must propose something usable. We have seen earlier how the very act of separation of use from making and designing creates the question of function; the need to generalize about use. When the patron is the user frequent interaction will solve the problem. When this is not the case the issue of function must be addressed, and chances are that it is put to the designer to do that unless the client does so.

Our simple linear model where form finds its way through subsequent transformations, from image to appearance, to object, to object transformed in use, can now be interpreted in two ways. If all the stages are distinguished but all are performed by the same party we have simply a model of a creative process where rapid oscillation of attention between various stages may take place easily. If, however, each stage is controlled by a different party, each exercising a specific mode of transformation during its moment of control, we suddenly have a new and complicated game. each step in the form's progress becomes a transfer of control. At that point the process is no longer just creative and technical, but acquires an organizational, if not a plain political dimension. In the transformations of the form we see a configuration of social relations emerge.

The question may be raised here as to what is the signature of the control we call designing. If we can recognize the dominance of use in the extremely complex but ordered forms of traditional Muslim settlements and if the dominance of making may turn into manufacturing or even into autonomous art, how is the dominance of design control to be recognized? It seems hard to find a form in which making and using fall so far behind that designing makes its own. While the designer will always make a difference, designing does not live by itself but by what it is placed between. If its task is indeed to transform a borrowed image into a representation, then design will reflect, like a mirror, whatever light reaches it - the dreams of the patron, those of the maker, those of the user or those of the person who is called the designer. And we may seek in vain for its own imprint.

HIERARCHY IN FORM

We have seen how a form, at any stage of its journey from image to object, may also find itself in a hierarchic organization. The engine is placed in the car and attached to the body frame. The house is placed in the block which is enclosed by the network of roads. The floors of a building are held up by a steel structure. Each form finds its location in a higher-level form and, in turn, each form makes place for lower level forms. The engine-block holds the pistons and other parts, the building made of walls and floors holds partitioning and furniture.

I have earlier argued how levels in a form hierarchy are domains of intervention. Transformations in the complex form reveals its hierarchic structure. Within the city block houses can freely be taken down and others built without disturbing the road network. The engine can be taken out of the car and replaced with another one without rearranging the car as a whole.

But intervention, the agent of change in artifacts, is the exercise of control. The hierarchic structure of the man-made world is therefore a reflection of control patterns. A level in the form hierarchy offers a domain for action within the stable context of the higher-level form. The hierarchies in complex forms are also settings for patterns of control.

It is particularly in the concept of so called 'open systems' that the issue of control comes to play an important role in the assembly of forms. We speak of an open system when its various subsystems are sufficiently autonomous to be replaceable by systems of another make.

With the complex machine the control hierarchy is usually related to an assembly of separate parts by different parties and to questions of maintenance during the lifetime of the whole.

When we think, for instance, of an office building in which partitioning of different manufacturers can be used by different users we can speak of an open system. Most buildings are to a certain extent open systems. We usually can bring in new and different heating systems, alternative electric circuits, new ceilings or partitioning, and additional plumbing.

There is a direct link between the concept of open systems and the issue of control because it means that a subsystem can be manufactured by anyone who accepts the constraints of the particular systemic hierarchy of the whole. When a system is open it is no longer necessary for a single party to control all levels of the complex form. Opening a composite system that one is in control of, always implies delegation of initiative to other parties. Often, our first impulse is to seek control on all levels. This is a human reflex but eventually, with complex forms, we learn that it is to our advantage to allow initiatives on separate levels. There may be a loss of control of details, but there is a significant increase in variability and a better chance for partial improvement without a need for overhaul of the entire complex form.

To have control by different parties on different levels, the interface between them must be formalized. Only when on each level the constraints are stable will it be attractive for new parties to take initiative. In this way, automobile tires, audio tapes and gramophone records as well as thousands of other parts in manufacturing became normalized. It is the normalized interfaces that reveal the structure of the composite system and which settles the levels of intervention. The result is meaningful variation that is universally applicable.

We find the same principles work in environmental form: the room within which furniture can be controlled, the office floor on which partitioning can be freely arranged, the city block within which buildings can come and go, the districts in the larger urban infrastructure that can change their road patterns with a fair degree of independence are all examples of the congruence of form levels and control levels.

As human beings, we see ourselves first of all in a web of human relations. We tend to interpret things around us as reflections of a social fabric and, in a broader sense, this may be correct. Where the understanding and representation of form is our subject, however, we do well to take a different perspective and must try to see how relations of controllers are defined by the forms they control. That things determine the roles played by those who exercise control is particularly evident in environmental forms. The street may be under control of the municipality that is represented by an anonymous bureaucracy, or it may be owned by an individual land owner, or it may be under control of the collective of those who live in the houses alongside of it. In all cases there is a party that controls the higher level street form which plays a role distinct from that of the parties that are in control the lower level house forms. The identity of controlling parties may change but the roles played remain the same. We must therefore distinguish the players from the game. The game is set by the levels in the complex form. On each level the identity of the controller may change. We play the role determined by the control that we exercise, by the level that we operate on.

Where each level in the hierarchic form can have its own controlling party, it is also possible that one party controls several levels. Thus, the office building with its changeable partitioning can accommodate a number of inhabitants in control of their lower- level systems while the whole building is managed by a third party. Yet that same building may well be in the hands of one single company that does not allow its divisions to exercise physical control on their own floors. In that case local movements of the partitioning may still take place but only when central management agrees to respond to local needs. Still, the distinction between different realms of control remains.

In the same way the student who rents a room from a landlady may bring in and arrange his own furniture which makes for a lower level configuration under his control. But when the landlady keeps one room for her own use the lower level configuration in it is under control of the same party that controls the higher level. It is not difficult to find other examples of various ways in which control can be distributed in different ways within the same hierarchy of form. The players may come and go but the play remains the same This a-symmetry between the stable control pattern determined by the form on the one hand, and the variable identity of the controlling parties on the other hand, gives the hierarchic form its autonomy.

Only on a very different time scale hierarchies themselves do change. The adaptable partitioning in the office building, for instance, evolved in the early twentieth century from the large building with fixed rooms. Under pressure of changing usage and suggested by new technology a new level came into its own.

HIERARCHY IN SPACE; TERRITORY

So far, we have discussed the control over physical forms, but our control in the environment is not limited to solid things. A space can be a unit of control as well. We are in control of a space when we can determine what things can be admitted and who can enter it. The control of a space amounts to the control of its boundaries. There are 'gates' that allow boundary crossing only by permission of the power in control. Such units of spatial control we can call territories. This definition seems to be compatible with the normal usage of that term.

The territorial power can keep things out and determines what goes in. When, for instance, the state of Massachusetts prohibits firearms within its boundaries it exercises territorial power. The landlord can forbid boarders to bring dogs into the house because he controls a territory.

As with any spatial unit the territorial space can be indicated by means of physical things. The gate, as an instrument of territorial control, is at the same time its expression. Other forms are also used to signal the extent of a territory. Sometimes they are functional like walls and fences; often they are symbolic like boundary stones or written text indicating to the traveler that he is about to enter a new territory. To really judge the extent of a territory we must not only watch the boundary's forms but also the movement across them. We must check the verity of the gates and the boundary signs; they may be remnants of bygone days and nobody may be there, but when we enter, we will find out. A territory is truly spatial, and physical elements do not by themselves make a territory. The suburban house, to give another example, is a closed form and it may have a strong gate- like door, but its territorial boundary lies where the lawn meets the sidewalk. Alternatively, a space surrounded by a fence may only serve to keep chickens in without having any territorial function.[26]

Hence, it is only the act of control and the presence of a controlling party that defines a territory. Territories without any physical boundaries are as real as those surrounded by solid stone walls. We find evidence of this when bathers gather on a beach: physical boundaries are negligible, but children and dogs are called back when they cross invisible lines and stray beach balls are retrieved with appropriate apologies. The invisible division of control is firmly established by people's behavior in the unmarked space.

All settlement is based on territorial order and it is not possible to understand the structure of cities, towns, and villages without a grasp of their territorial organization. Although good designers of environmental form possess an innate understanding of it, territorial control patterns are seldom explicitly discussed. We are no longer interested in the architectural expression of spatial control, leaving its function to technical instruments. Perhaps our neglect of the concept in design theory so far is because territory is not space determined by forms but by control. It is therefore, in the eyes of those who design, an elusive entity. One cannot design territory because its true deployment depends on the patterns of settlement that actually take place in the form. However, each form frames, by its spatial organization and hierarchy, the territorial options available to the inhabitants. The designer will do well to assess which territorial patterns are facilitated by his forms and which are discouraged by them.

Knowledge of territorial order is not only of interest to the designer of settlement forms. As I hope to show in more detail later on, it is useful in other cases where complex forms must be designed and made. Indeed, it seems that wherever space is shared there appears territorial structure When design activities are distributed among different designers within a shared larger space a territorial order will establish itself if only for the duration of the design process.

When complex forms like ships, planes, and buildings and other technically intricate assemblies are made, the distribution of design control among various designers and the avoidance of conflict among them can pose problems. A good deal of effort may be spent in finding out which parties may place their elements in what space. Such problems are familiar enough, but they are not normally seen as territorial. We might do well looking into them from the perspective of territorial organization.

A territory, once established, gives us a space to use at our discretion. Being in control we are free within its boundaries to arrange forms as we please. The extent of our own territory also makes us know neighboring territories as well as the public space in which are found the shared facilities, we collectively can make use of. This is true for the citizen who inhabits a house along the public street, but it seems equally valid for all controllers of form who must operate in a limited space shared with others, be this shared space a town or a machine.

Therefore, although the idea of territory is firmly rooted in environmental use, we may benefit from it in the design of complex forms in general. It may well become an issue of importance in general design methodology. For this reason it seems of interest to examine the concept somewhat closer.

TERRITORIAL HIERARCHY

The typical territorial hierarchy is one of inclusion: the municipality finds itself in the state, the district is in the town, the house is in the district and the private rooms are in the house. Each territory is located in another, larger one.

Let us once more consider the example we have used before, that of the resident landlord who rents out rooms. He may allow or disallow a renter to bring a piano in her room but does not control the piano. The objects that he personally controls can be found in spaces accessible to all occupants, like the stairway and the hallway that both he and the tenants use, and, also, his own private spaces where renters have no access.

We find how in the larger territory two kinds of spaces can be distinguished. There are the included territories controlled by the inhabitants in their rooms and there is the space accessible to all: the hallway, the stairs, the corridor and possibly a common room. The former are private spaces relative to the latter. This division of the larger territory in two kinds of space - the private space that is the sum of the included territories and the public space shared by all - is common to all territories on all scales.

We see here the makings of a territorial hierarchy based on the principle of inclusion. In their distribution of control we find properties that are rather similar to that in the form hierarchies that we discussed earlier. The division of a territory in a public and private area, for instance, shows how a collective of private spaces, themselves being separate territories, needs the addition of a public space to make a larger territory. This is similar to the way in which the leaves need a twig to make a branch and the furniture needs walls to make a room. Moreover, both the territorial hierarchy and the hierarchies of physical configurations are determined by control: they both are 'control hierarchies' accommodating controllers on different levels.

We also see, in the example of the landlord, how one party can be in control of the larger territory and at the same time control one of the included territories The landlord has his own private space which is not accessible to the renters but is, like the renter's room, an included territory connected to the public spaces in the house. Hence, we find here, as we found in form hierarchies, a certain autonomy in the order itself. Control units must be distinguished from the parties exercising the control. The spaces can be there, delineated and bounded, while the identity of the controllers may change.

At the same time, territorial hierarchy can be more fluid in its spatial structure. Because it rests primarily on the control of space it can change easily and let physical form follow.

We may again examine the example of the landlord as illustrated in *Figure 4.3*: the same room can be a) a separate territory controlled by a renter or, b) part of the private territory of the landlord or, c) when converted into a common room, part of the public space of the larger territory. In this example the physical form is always the same but the territorial order changes with the control pattern.

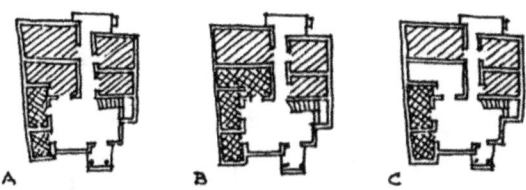

Fig. 4.3. Same plan of a house with a) the same room middle left being rented out (light hatching), b) being part of the landlords' private space (dark hatching), c) being part of the communal public space.

To give another example: The house, first built and occupied by a single family, later has been divided in two apartments. In that case a new public space is needed for the larger territory in which the two apartments are included territories. We may find the two apartments share a driveway, some parking space, a front door and a hallway. Within each of these apartments we will find, as usual, rooms that are private, included territories that share the apartment's 'public' living spaces. The overall territorial depth of the entire house has increased because now there are three levels where at first there were two. But this particular change of control needs some physical change too; inside the hallway we may find doors to two apartments, one upstairs and the other on the ground floor, and outside there may be a boundary form - a hedge or a low wall - separating the shared parking space from private garden that belongs to the downstairs residents. (*Fig. 4.4*)

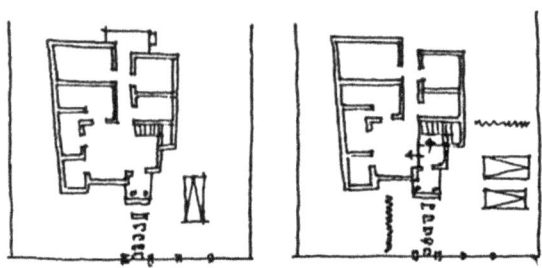

Fig. 4.4. Same house plan as in fig. 4.3; to the left and, to the right subdivided into two separate apartments with a common entryway leading to two territories

LEARNING FROM SETTLEMENT FORMS

What has been said so far may be sufficient to demonstrate how our patterns of control run on two distinct but interrelated tracks. With each intervention that we make in the physical world we can relate to other controllers in two separate systems. We relate in a hierarchy of form where we meet parties operating on higher or lower levels of intervention than ours. And we relate to those in control of territorial space with whom we share boundaries. When, for instance, we build a house a single intervention brings us in relation with a number of other parties at the same time.

This situation, as a general condition for environmental intervention, may seem complicated on first sight. But when we think about it we will conclude that those relations among controllers are familiar to us. When I build my own house I do so within a territory within which I can act. Within rules and regulations set for all house builders, I am free to determine what to do. The form produced by my intervention connects to the outer, larger, world in precise, predetermined ways: For access, sewage, water, gas, and electric power the systems in my house connect to higher level systems found outside my territory.

The territorial order is particularly helpful to allow the continuous process called inhabitation to happen. However, sharing space is not limited to environmental design. Wherever we must act in concert to make complex forms, a similar situation seems to present itself as well. If we think, for instance, of design for what goes inside an airplane fuselage or inside a ship's hull we can see how distinct design tasks relate to one another in the two ways discussed above. There are technical interfaces where physical systems must connect to a higher-level system. A discreet volume is available for each space unit and each technical sub-system, subject to negotiation with abutting parties. There are also public 'channels' in which cables run that feed lower level systems in locations where use happens.

The patterns of inhabitation to be found in urban environments, where continuous intervention over a long time has created a balanced and coherent integration of territory and form, may well provide useful models for the structuring of design processes for all complex forms. The common question is how different design tasks can be performed simultaneously with a minimum of interference while each has its own location in a larger shared space. Levels of intervention structure the distribution of design tasks for physical entities. Territorial structure allows for simultaneous intervention in a limited shared space.[27]

NEGOTIATION

To illustrate the uses of control models in the design of complex forms we can take a simplified scenario, but one complete enough to embody most common relations. Suppose there is a designer DA who is working for a client CA who owns and inhabits a piece of land LA. The client wants to accommodate three families on his land along with his own house and agrees that they should have their own houses and gardens there. These three parties CB1, CB2, and CB3, have very different life styles, needs, and expectations. Client CA states he wants a large part of his land available for communal use. He also wants designer DA to take care of all his relations external to his property: access, sewage, set-back rules, and such. Finally he states that speed is important. The faster the solution is found the better. It is clear that DA must decide the extent and location of the territories in the larger one controlled by CA, and that CA expects him to take care of the public space in his territory as well. Thus DA operates on two territorial levels at the same time.

Fig 4.5 gives a diagrammatic view of the territorial arrangement in this example: neither sizes nor locations of the three territories TB are as yet decided within TA. One way for DA to go about it is to first talk to each of the families to find out about their specific needs and preferences. Thus DA studies with great care each of the three sub-problems and comes up with three possible schemes that satisfy the individual parties CB. However, when he tries to arrange these schemes in LA he finds it difficult and feels that the land is not utilized well. He also has problems to accommodate the desired arrangements to the external conditions. He goes back to the three sub-schemes to reconsider them in light of what he has learned.

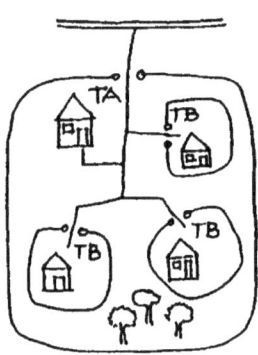

Fig. 4.5, Main territory TA with owner's house and three included territories.

Another approach would have been for DA to first study the external conditions carefully, talk to CA and then make a quick, educated, guess to allocate land for each of the three houses and perhaps even make a rough sketch of what they could be like. With those first ideas he approaches the three families and finds out, to his dismay, that his assumptions do not work and, in fact, trigger the ire of parties CB who dismiss his ideas and demand that he listens to them first.

Client CA, however, is happy with the scheme because it follows his suggestions. Nevertheless, since the programs of the three houses must be met, he agrees that DA works with the users some more and to then come back to the larger scheme.

In both scenarios Da's attention ultimately moves up and down between higher and lower level issues. In the first case he starts from the bottom and in the second he goes from the top down.

Now, finding himself in an elaborate iterative process DA decides that he cannot put enough hours into the job himself and brings together a design team from his office. In the beginning the team has lengthy discussions about whether to begin at the bottom or the top. When they finally decide one way or another, they go through the same different steps DA already followed. DA, however, wants to stay in control of things and insists on making all the final decisions. Team members may talk to the three users CB but cannot make decisions on the spot; they must bring their proposals in the design team.

Soon, it becomes clear that this procedure does not speed up things. It may relieve DA from time consuming work but the cycles up and down the hierarchy are the same as when he would work alone, whereas in this case his insistence to make final decisions is making the process more elaborate. Someone proposes to buy a computer program to speed up parts of the work...

At this point client CA as well as his customers CB become impatient with what is going on. Each feels that DA does not represent their particular needs and gives too much attention to the interests of others. Designer DA is unhappy too because he finds himself caught between four parties while at the same time trying to keep his team going and making sure they do things that he can justify to his clients. Team members are frustrated because they feel that their more detailed knowledge is not recognized, and they become tired trying to get their point of view across to the boss.

Finally client CA acts and decides that his users CB will have their own designers DB1, DB2, and DB3 to work for them while DA will work for him. These four designers must come to a joint decision and a premium is set for any gain in time made.

While DA studies alternative schemes for allocation of the territories of CB1, CB2, and CB3 in relation to external constraints, seeking maximum use of the available land, the other three designers work at the same time with their clients to study alternatives that maximize use of their territories. The four meet regularly to coordinate their work. Soon an understanding of interfaces is worked out about issues where interests of A, operating on the higher level, confront interests of the three B's.

We have a different team where single designers identify with territorial responsibilities which are the same as those that will operate after completion of the job. Design responsibilities now follow the control pattern implied by the subject. In contrast with the first approach, territory is now structuring the relations among designers and it is clear who is deciding individually about what and what is to be decided upon jointly.

No designer DB can come to final decisions unless he is given a first general idea of the space available to him. In fact these territories TB are now elements with which DA works. He cannot work within the territories DB but he can negotiate about their extent and their position in the larger scheme. He must also make all decisions relative to what goes in the public space - that is to say the space not occupied by the three included territories TB. He provides the infrastructure to which the designers DB's relate.

It is of interest to note that in this model the actual size and disposition of the territories TB in the larger territory TA can be undetermined in the beginning. The group of the four designers may indeed negotiate about such things as well as about the provisions needed in TA's public space to serve TB. The latter have to do with the connection of the utilities in TB to those in TA (interface conditions) and also additional rules pertaining to all three forms in TB such as, for instance, uses of materials, colors, building height, even typology, to enhance the coherence of the overall scheme.

This hypothetical and inevitably simplistic example may illustrate how a model of negotiation may be part of a structured design process. Everyone who has ever operated in the design world knows that such negotiations are part of the game. However, they are seldom seen as efficient ingredients, positive to the design process. The prevailing, but implicit, theory is that the job should be just technical.

Working in a territorial structure can facilitate the coordination within a design team. The problem in the example is rather transparent and would perhaps, in real life, not need such a formal organization. But once the forms become more complex, and the number of designers that are involved increases, the coordination of design activities becomes a serious issue. In that case, explicit acceptance of hierarchies of control, related to form hierarchies and territorial hierarchies will help. Indeed we have here the ingredients for a general model for the distribution of design responsibilities in complex form making.

The link between explicit control patterns and design is particularly alien to technical design thinking. The idea of territorial negotiation suggests that our traditional ways of evaluating and modeling things are not efficient by themselves. The process of territoriality and control among designers cannot be captured in objective terms. Being socio-political in nature it cannot be computed and therefore its results are not entirely predictable.

A formal approach in design management based on control patterns may initially create ideological difficulties in design circles. But such difficulties, important and real as they are, will eventually be overcome when we find that the control-oriented approach which is suggested here leads to better results with less effort. Whether this is truly the case will, of course, be a matter of dispute for some time to come.

LEARNING FROM SETTLEMENT FORMS II

A good argument for territorial organization in design of complex forms is found in the challenge posed by the layout design for very large-scale integrated systems in computer technology.

The layout design for a VLSI system is part of a sequence of presentations needed to produce a 'chip.' According to one witness there are, in broad outlines, at least four appearances needed between the initial idea, as expressed in a first memo or perhaps verbally, on the one hand, and the actual printing of the circuit on the other hand. These could be called 'Behavioral Definition,' 'Analog Circuit,' 'Digital Circuit,' and 'Layout.' Each of these terms denote a group of 'appearances' within a specific language of representation. Each such group entails a complex design process by itself.[28] The Layout, for instance,

will consist of various kinds of 'maps.' The documents that eventually reach the producer are called 'Masks.' These are extremely accurate representations of each of the networks that must be traced, in various techniques with various materials, on the surface of the small silicon chip, overlapping and complementing one another to make a working circuit. To the manufacturer of the chip the Masks function like templates to etch the circuit on the silicon wafer.

The layout design must produce the actual distribution of the circuit parts on the silicon surface. It is the result of decisions about actual physical elements and their positions relative to one another in space. The layout must answer conditions spelled out in the other appearances of the form; the digital circuit and the analogue circuit. At the moment of this writing it turned out that in many cases the layout design was the critical path of the entire development, its duration proving impossible to predict and sometimes causing serious delay of the delivery of the product in which the chip was to function.

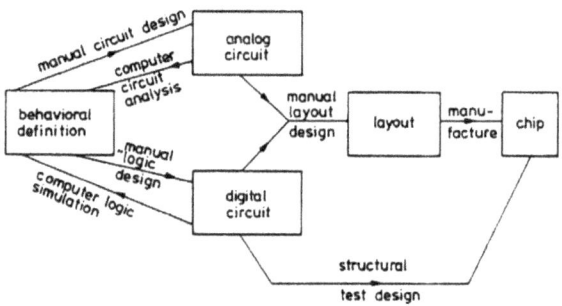

Fig. 4.8, Traditional VLSI system design procedure. It demonstrates a string of appearances with parallel branches as introduced in Essay 2. It also demonstrates the 'leftward' and 'rightward' moves that are possible in any string. Characteristically the leftward moves are systematic evaluations that therefore can be done by computer simulation, while the rightward moves are the leap toward the form that need another methodology and therefore are labelled 'manual.'

The smallest functional element in a VLSI is the 'gate,' a junction between two or more conduits that channel electric currents. A single VLSI today (1983) may have a half million gates and the number of them that can be squeezed on a single chip is rapidly increasing.[29] The primary task in layout design, to put it simply, is to distribute and connect all these primitive elements on the available surface in such a way that the required performance is met.

To make this task at all manageable the other representations (the analogue and digital circuit design) serve as a program for the layout designer. They also spell out a hierarchy in the ordering of the gates: gates make 'transistors,' transistors make 'cells,' cells make 'macro's' and macro's make 'block

according to one account.[30] The hierarchy that is thus

established is similar to the control hierarchies discussed earlier. The elements on one level are connected to one another by means of additional parts. In other words, there is always a higher-level form to which lower level configurations are connected or, to say it from another perspective the lower level forms connect to a shared 'infrastructure.'

Here, available space comes into play. If we see the configurations that constitute a level as units of space, we see that the connecting parts always must lay outside these 'private' places: they are found in a 'public' space; public that is to the forms that connect to them. This pattern repeats itself, in true hierarchic fashion, on all levels of the form.

It is this territorial quality which is of interest here. Indeed the VLSI system is an almost perfect example of a territorial structure. This suggests the possibility of a distribution of design tasks among different designers who each have a clearly defined territory within which they must design and manage the division between 'public' and included spaces in it. Designers control their own included territories and connect to higher level forms the use of which they share with other designers operating on the same level.

The analogy with settlement design is obvious. Limited space must be shared to simultaneously design a number of forms on different levels. This analogy suggests opportunity for design efficiency where complex forms are involved. However, this approach demands that methods of territorial negotiation be adopted in an engineer's culture in which the elimination of the human factor in the design process is considered ideal. Territorial negotiation seeks to utilize the interaction between designers to find the best possible form that allows the most compact layout. It recognizes the social dimension of a design process, a dimension instinctively distrusted by the engineering profession.

DIFFERENCES

Settlement forms and VLSI layouts both are hierarchically structured complex forms deployed on a limited surface according to fairly strict rules of access and proximity. In a town, access means people and vehicles and goods moving in and out of the various places. In the integrated circuit this means electrical currents and therefore connections by circuit paths. Distance in the circuit layout influences the time it takes for an impulse to travel from one place to another. Proximity is extremely important and in an ideal layout two units having frequent interaction lie close to one another. Beyond such similarities the differences are many and profound. Among those of interest to us here are the following:

The VSLI circuit is homogeneous. It is like a city in which everything, from furniture to the major road network and all buildings, partitioning, sewage systems, gas pipes, water distribution, alleys, roads, roofs and floors are made out of brick. Or, rather, where we see only one technical system. This homogeneity of the complex form was somewhat approximated in traditional history mud-brick towns of the Middle East and we have seen previously how extremely complex those settlement forms could be. But the contemporary town is not homogeneous in its materials and systems. It is a combination of many different technical systems that have their own components made from a variety of materials and their own hierarchies. The VLSI is basically a hierarchy of one technical system performing all expected functions.

Moreover, the VSLI layout is, in all its complexity, a machine, albeit perhaps one that seems to test the wits of its makers. As all machines it is, as we have seen, to be observed from the outside and to be judged from there as to its performance. The town, being a settlement form, is to be judged from the inside and has no performance towards the outside for its major purpose of being there. It is built to allow those who inhabit it to perform their daily lives.

We examined this fundamental difference between the 'inside' and the 'outside' view in the previous essay. Because we inhabit the urban form its parts are in continuous control of different parties and, therefore, in continuous partial transformation over time. A human settlement is never finished but always in a state of building and design. The circuit, on the other hand, must be finished and, once the design is completed, it is grafted in silicon never to change again.

It is this difference in the available time frame that we must keep in mind when we seek an analogy between urban forms and VLSI's. The town, in its greater complexity and inhomogeneity and with its wide range of criteria, is nevertheless possible and ordered because it is a living organism which can start small, grow over time and adapt continuously to new uses and circumstances. Most importantly, mistakes can, and are, constantly corrected. As humanity's most complex and baffling artifact, the cities we make have the grace that they can, like plants and living bodies, recuperate, grow, be sick and become healthy again. Their processes of design, use, and making, and therefore their patterns of control, never stop operating.

But the layout design team must arrive at its complex form in the shortest possible time. The VLSI system compares to the settlement form only in the period of its design. In this stage design control patterns seem similar to those in settlement forms. The designer of the VSLI layout are its inhabitants, but only during the design process.

MATCHING HIERARCHIES I

The hierarchy in the layout of the Very Large-Scale Integrated system is of course not established by the layout designers. It must match the hierarchy of the Analog Circuit and the Digital Circuit. In these earlier appearances of the same form the hierarchy has the same purpose; to identify functional units that are, at the same time, control units in the design stage. Their vertical distinction is, as we have seen, visible by their transformations. The lower level forms can change freely within the constraints set by the higher level.

In all design processes of complex forms it is important that hierarchies in two consecutive appearances match. The earlier appearance can only effectively serve as the brief for the next one if this is the case, so that the next one can be tested against the criteria implied by the first. However, this matching need not be a complete mirroring of hierarchic structure. It may well be, for instance, that one level in an appearance will be translated in a two-level form in the next one.

For example we may, in a first design of a facade, represent a window as a single form with a frame and its subdivision in panes and mullions. In the next representation, where the actual technical decisions are made and profiles are determined, we may find it convenient to distinguish a window frame in which alternative panels can be fitted in. A fixed panel can be replaced by one that can be opened, a transparent one can be replaced by one that is opaque. If that is the case, the window is a two-level organization. This may simply reflect the designer's desire to determine the exact panels at a later date, or to technically adjust the same window frame to various locations and uses in a building.

Once the distinction in levels is made, however, it may go with a division of control as is often the case with form hierarchies. Staying with the example of the window frame, we can think of the architect designing such frames but leaving the decisions about the panels to the inhabitants of the building, as has been done in the design for adaptable dwellings. In this case, the technical principles are not new. The innovation is that the architect used the hierarchy as a basis for the delegation of control.[31]

Fig. 4.6. Left, small territories and a single large one all accessed from the major roads. Right, a deeper territorial hierarchy with the large territory holding many included territories with a new road which is public to the newly included territories but private relative to the major roads.

The reverse can happen as well. It may be that in one earlier appearance two levels are distinguished while in the next their joint performance is seen as a single level under control of only one party. An example can be the case where an urban designer suggests a residential street giving access to a number of houses, while in a next stage the design of the houses comes under control of a single party that sees the whole as one project. This may produce a proposal for a single apartment building while no public street is needed any longer, nor is there a further lot division. (see *Fig.4.6*)

The merging of levels often occurs when one party takes control of all units on a particular level. Obviously, this entails a negation of the intentions of the previous design party and negotiation may well be in order. As a general rule, however, we may assume that hierarchic depth will increase as the form progresses in its transformations from idea to object.

In all such cases, whether the hierarchy intensifies or flattens out, alternative control patterns are involved. A change of form hierarchy always has its implications in terms of control opportunities.

MATCHING HIERARCHIES II

So far, we have considered hierarchic forms in their transformations through various appearances and we discussed the matching of the hierarchies in these appearances. Doing so we had in mind the model of the design task found in the previous essay:

I [HA1 [HA2 [....HAn [O

where I is the initial image and O the artifact to be produced, while HA is the hierarchy of a complex form, expressed for different purposes in different ways; HA1, HA2, HAn, etc.

There are complex forms, however, that must be understood as combinations of a number of distinct vertical organizations. Once more, settlement forms give an example: The road network can be seen as one form hierarchy, and the sewage network, the water distribution and the gas distribution as others. They each have their own structure complete with different elements and different controllers on their various levels. Each is distributed across the same area. Their sum makes a coordinated infrastructure of urban utilities.

We may find a similar combination of hierarchic forms in the large office building. Here we have not only the load bearing framework that is deployed through the entire complex form, but also the heating, ventilation and air-conditioning systems, the fire prevention system of sprinklers, and the electric and telephone circuits that all are deployed through the entire building. They each are the product of their own designer expert. They each make a hierarchic form and assume different controllers on different levels, if not in the design stage then certainly in their use and maintenance. They must, eventually, match to make a coherent building. The building is indeed the result of the merging of a number of form hierarchies into one complex whole.

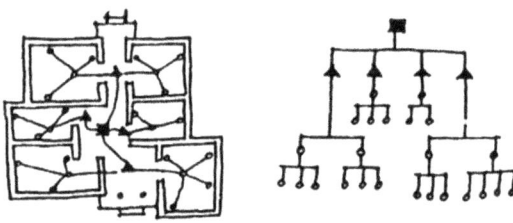

Fig. 4.7. A spatial organization combined with a hierarchical service system the diagram of which is shown at the right

Here as well, matching is important, but in this case, we do not deal with different appearances of the same form but with independent hierarchic forms that must combine in space. Once again, the matching need not be one on one. The electric circuit in a normal house is distributed by groups of outlets each protected by a fuse at the point of separation. One group is usually congruent with a number of rooms. We see how one unit of the group level in the circuit hierarchy relates to several units on the room level of the spatial hierarchy of the building. This circuit, however, knows one lower level, that of the outlets. Here the reverse is true: several outlets match with one room. Although the relation between the two hierarchies is clear their levels do not match one on one but in a kind of staggering way.

Once we begin to observe hierarchic forms, the ways in which they can be combined are many. In the previous example we found a hierarchy of spaces to be matched with a circuit hierarchy. This illustrates how the hierarchic systems that we see in the forms need not be limited to physical configurations but that there is always the reverse: the spaces formed by the placement of physical parts. Of course, in settlement forms space is the most important entity and has its own identity, but in most complex forms we find space made available by one physical system become the location for another physical system. Our thinking shifts easily and continuously between solid form and space form, but we must keep in mind that space, available by means of form is different from the controlled space we call territory. As we have seen earlier in the *Figures 4.3 and 4.4*, spaces offered by a physical form can be differently interpreted territorially.

We find a dialectic relationship. A territorial interpretation directs the distribution of subsystems to serve territorial units. Once the physical organization is in place, however, a different territorial interpretation is possible and, in turn, adaptation of physical systems may follow. And so on.

LAST LOOK AT THE MODEL

We now must return to the model of the design task and trace the path that we have travelled. Because, in the true spirit of our subject, this model, beginning as a simple string of appearances, has itself been transformed while we proceeded.

First, we had an expression representing a linear progression of appearances of the form each one setting the context for the next; each interpreting the possibilities and intentions given by the previous one. The form, in this view, grew from the first idea, only present in our imagination, to become, via progressively more concrete appearances, an object in the 'real' world of things and people:

I [A1 [A2 [...An { O

The object, thus born, finds itself to be part of the inexhaustible juxtapositions of the many things we live with. It continues to participate in people's affairs.

In that one-dimensional interpretation we found that each appearance is, in fact, a form in a hierarchic order. There is always a form on a higher level serving as context for the transformations at hand. Often our design is made in an already existing higher-level form:

A is included in HA

Next, we found that such a hierarchy, in which our design finds its place can be, itself, subject to the transformation from image to object. Complex forms constitute themselves hierarchically and proceed likewise from one representation to another.

I [AH1 [AH2 [.....Ahn [O

Finally we found that even more complex forms are, in fact, juxtapositions of different hierarchies, each with their own elements and establishing their own order, each proceeding likewise from image to object but also connecting laterally since they find their deployment in the same space, so that their juxtaposition results in a single object.

[HAa1 [HAa2 […..HAan [
I HAb1 [HAb2 […..HAbn [O
[HAc1 [HAc2 […..HAcn [

We can say that for each form at least three hierarchies may be seen in parallel: the physical form itself, the spaces made by the form, and the territorial interpretation of the form. But since the physical form can be several systems itself there can be more.

With this elaboration we have arrived at a three-dimensional diagram of the transformations of a truly complex form, examples of which we found in urban environments or large buildings. But we suspect that these properties are not confined to forms of habitation but may be inherent to all complex forms that are inhomogeneous and combine different systems.

Most importantly, we found how the model helps us to understand how, at each stage of the transformations that take place, and at each level of the hierarchies that are constituted, as well as in each hierarchy that is added to another, different parties may be in control and different expert designers, each with their own language and way of working, may operate. Indeed the model appears to be one of control units in the design process. This means that each participant in the creation of a form, at any stage, may find connection with other possible parties in six different directions: First to the left and the right in the design string, next up and down in the form hierarchy and, finally back and front towards other, matching form hierarchies.

And this is not the complete picture because each time that we meet another party in the design game we can only cooperate when we share some values and interests: these make for the thematic systems which we discussed in the third essay. And since thematic systems are not bound to time or space but only to social groups, they make us not only meet other parties in control of some form or space but also that party's social context.

This is not shown in the model but should not be forgotten.

THE SOURCE

Having thus completed our model we may ask ourselves to what advantage we have found in it, because it was not primarily my intention to develop such a construct. It came along because we explored the ways in which the design task rests on human interaction. Its merit is that it illustrates the many ways in which people connect to one another when engaged in the emergence of a single form. The model itself is perhaps already too intricate, which suggests that the argument has run its course and it is time to stop.

The kind of abstract elaborations that we arrived at are the inevitable result of sustained rational exploration but does not necessarily mean that the actual making of complex forms needs to be difficult. Complex forms have a way of coming about and operating smoothly whenever people act in concert and freedom: this much we found in the first essay.

When we do realize the many interactions our designing implies, how is it possible that forms come about at all? How can buildings get built, integrated circuits get manufactured, airplanes assembled and a thousand other equally elaborate things be created every day? Or, to be more precise, if we must coordinate and organize ourselves how can we have, when for instance a building is built, so many experts all working at different systems, each system constituting its own hierarchy, finding its own way from image to object and yet, somehow, being part of an even larger whole? How can this be in other complex forms? Indeed, how is coherence in planned complexity possible at all?

The model that we have developed may be of some help to coordinate professional activities, it may perhaps suggest part of what design is about but, in the way of such artificial constructs, it begs the question. Somehow all these parties can only operate, can only coordinate effectively if there is, beyond their expertise, beyond the languages that serve them in that expertise, a shared image that all can relate to: something implicitly understood by all that makes the whole become more than the sum of its parts. Unless participants share such a vision, the form must inevitable fall apart.

As we discussed in Essay One, complex and sophisticated forms come about when a tacit knowing of them is shared in a social body. We can safely assume that to be true today as it has always been in the past.

Designing emerges in a society when tacit understandings among makers and users are no longer sufficient. We then begin to discuss what to do and describe in words, drawings, or models, the forms we seek agreement about.

In that way, forced to move away from self-evident solutions we can come to unfamiliar forms of startling novelty utilizing new knowledge, new technology, new materials and new ways of seeing.

But our inventions and proposals must, to lead to coordinated action and a final physical form, be shared in a social body. Like flying fish emerging from the dark waves to glitter for a moment in the sunlight, the designed novelties, visible in the stark light of explicit discussion must eventually become part of a living culture, and disappear in the vast realm of the self-evident - the domain of silent understandings that are unquestioned and secure; the domain from where designing rose in the first place.

THE END

ILLUSTRATIONS

First page, Chapter 1. Canal houses in Amsterdam showing shared patterns and typology

Fig. 1.1. Part of the plan of the town of Pompeii from the map by Overbeck, see note 9. Four house plans have been shaded to illustrate the wide range of sizes within a same type.

Fig.1.2. 'Botter,' Traditional flat bottom fishing ship. After Peereboom, see note 11.

Fig.1.3. Plan of an historic house type in Medina, Saudi Arabia. The plan shows an entrance with direct access to the space where the owner conducts business, then lea n). The route ends in the main hall where guests are received which is the full height of the building with a small skylight. The central part is flanked on two sides by raised floors enabling host and guests to sit on cushions. After Kasjughee. SMarchS Thesis, MIT.

Fig.1.4. Various plans of the traditional Malayan house type with the house floor raised from the ground and accessible via the 'Serambi' - a porch like space.

Fig. 1.5. Plan of the town of Economy, Pennsylvania, 1876, Showing typical gridded street layout with variable occupancy of buildings all at the edges of the blocks as was the custom. The divisions of property within the block not shown. From: Joseph A. Caldwell, Caldwell's Illustrated Historical Centennial Atlas of Beaver County, Pennsylvania. Conduit, Ohio 1876. After John W. Rep, see note 14.

Fig. 1.6. Part of map of historic Cairo, showing lack of predetermined geometry in the street pattern, but with a clear hierarchy of major streets, secondary streets, and dead-end streets. The private courtyards of the houses are shown as well, so are the boundaries of the individual houses. After: Sawsan Bakr, SMarchS Thesis, MIT.

First page, Chapter 2. Tony Garnier; La Cité Industrielle, 1901-1917, detail

Fig. 2.1, various appearances of the same form, chapter 4

Fig. 2.2, constraints to form relations, chapter 10

Fig. 2.3, capacity exploration, chapter 12

Fig. 2.4, solution space, chapter 12

First page, chapter 3. Detail of the main portal of the church of Sainte Foy, 11th century. Conques, France.

Fig. 3.1 Detail of map of Venice, by Barbari, 1539.

Fig. 3.2 Hierarchy of street systems.

Fig..3.3 Tree system hierarchy.

Fig. 3.4 Two levels of intervention.

Fig. 3.5 Three level hierarchy, street, lots, houses.

First page, Chapter 4, Garden boundaries, Commonwealth Avenue, Boston, USA.

Fig. 4.1. Part of the excavation of the city of UR, ca. 2000BC, after Sir Leonard Woolley.

Fig. 4.2. Part of Tunis Medina, dead end alleys shared by clusters of courtyard houses. After the base map of the Association Sauvegarde de la Medina, Tunis.

Fig. 4.3. Same plan of a house with: a) the same room, middle left, being rented out (light hatching); b) being part of the landlords private space (dark hatching); c) being part of the communal public space (white).

Fig. 4.4. Same house plan as in fig.4.3 to the left and, to the right subdivided in two separate apartments with common entryway leading to two territories.

Fig. 4.5. Main territory TA including owner's house with included territories TB.

Fig. 4.6. Left, small territories and a single large one, all accessed from the major roads. Right, a deeper territorial hierarchy with the large territory subdivided holding many included territories with a new road - which is public – leading to the newly included territories, but private to the major roads.

Fig. 4.7. A spatial organization combined with a hierarchical service system the diagram of which is shown at the right.

ENDNOTES

[1] David H. Friedman, Florentine New Towns, Urban Design in the Late Middle Ages. The Architectural History Foundation, New York and MIT press. 1988. Friedman shows how in the Florentine new towns the geometric lay-out constructions begin to depart from the rote surveyor's grid, thus suggesting a development from 'organic growth' via laid-out towns to towns "designed" by professionals. His evidence tends to correct the general notion among historians that new towns at that time knew no urban design at all. Thus he pushes forward the point of emergence of the professional designer in town design in European history.

[2] John A. McPhee, The Survival of the Bark Canoe. New York, Farrar, Straus and Giroux, 1975. The book includes "A portfolio of sketches and models of Edwin Tappan Adney (1868-1950)"

[3] Andrew Sheen Papuan Canoes, in I.F.D.A. Dossier 40, March/April 1984. Nyon, Switzerland. "Some families, who live right on the beach under the coconut palms, specialize in canoe making; whilst families who live up beyond the mangrove swamps made clay pots, and weaved the sleeping mats from pandanus leaves. A system of barter still prevails with fifteen or twenty clay pots procuring a family size canoe."

[4] Michael Lenehan, "The Quality of the Instrument, Building Steinway Grand Piano K 2571." In the Atlantic monthly, August 1982. One part of the long article is titled: "Searching for the Master Plan." It begins with the following sentence: "As I followed the progress of K2571 through the factory, it gradually dawned on me that the people I was watching were assembling a very difficult jigsaw puzzle - one made up of tiny but critical details - without ever referring to the complete picture. Some of these details seemed to exist only in the heads of the workers who needed to know them."

Eventually, the writer is introduced to the pattern shop where the old man in charge there explains: "the chief engineer....had his own book and he just wrote down everything he ever saw, how it was done, how it was made. And every foreman had a little book....in his back pocket.... It was everything more or less verbal - passed down from son to son, as far as the Steinway family was concerned, and from foreman to foreman. There was no such thing as specifications. There were some drawings, and where there weren't drawings they had patterns. And that is why it was called the pattern shop. The major pattern was hanging on the wall or, if it was small, in some drawer...if...something happened to the secondary pattern which was in the factory, okay, then they came here, and we compared, or made a new one from this, et cetera. But there were very few drawings as we know them today."

Further on the point is brought home: "I asked....if I was understanding this correctly. Was he saying that the model D, for example, was not made from drawings and specifications? That the drawings and specifications had, in effect, been made from the piano? He walked to one o the tall metal cabinets. "The only D scale drawings we have passed on to us," he answered, "are from finished piano's after the fact..."

[5] For a more detailed survey of informal construction methods in Cairo see also: "The Housing Construction Industry in Egypt. p.63-91. Comparison of three low income environments." Joint Research Team Cairo University / Massachusetts Institute of Technology: Interim report, phase 1, 1978 Technology Adaptation Program MIT/TP Report 78-3, Spring 1978.

[6] For a more detailed survey of informal construction methods in Cairo see also: "The Housing Construction Industry in Egypt. p.63-91. Comparison of three low income environments." Joint Research Team Cairo University / Massachusetts Institute of Technology: Interim report, phase 1, 1978 Technology Adaptation Program MIT/TP Report 78-3, Spring 1978.

[7] The term 'informal', in urban residential construction is generally used for those activities that take place outside the official plans and projections and without required permits. It is not only the urban poor who build their own this way, but it includes many others who aspire to owning a house to live in. Land used for these constructions is not always obtained illegally by invasion but may be purchased from landowners who subdivide and sell their land in disregard of government planning. By political pressure or tacit agreements

informal neighborhoods over time obtain municipal infrastructures like road pavement, electricity, water and sewer networks, often after people's factual control over their property has been legalized. When successful, an informal development produces an urban environment within two decades that cannot be distinguished from any other user controlled low-rise high-density urban environment.

8 I am grateful to Architect Alphonso Govela for presenting me with a copy of the Tolteca manual and explaining the circumstances of its genesis.

9 For a more detailed survey of informal construction methods in Cairo see also: "The Housing Construction Industry in Egypt. p.63-91. Comparison of three low income environments." Joint Research Team Cairo University / Massachusetts Institute of Technology: Interim report, phase 1, 1978, Technology Adaptation Program MIT/TP Report 78-3, Spring 1978.

10 For a more detailed survey of informal construction methods in Cairo see also: "The Housing Construction Industry in Egypt. p.63-91. Comparison of three low income environments." Joint Research Team Cairo University / Massachusetts Institute of Technology: Interim report, phase 1, 1978, Technology Adaptation Program MIT/TP Report 78-3, Spring 1978.

11 When discussing house types in my class, I asked students to describe various types on the basis of the documentation made available to them. It was an excellent experience for all of us to learn how varied and rich the information implied in a type can be and how any description is a reduction which only makes it more difficult to agree. I became convinced that the way to learn to know a type is to work on it under the guidance of others who already know it. Acquiring such working knowledge comes naturally.

12 John W. Reps, "Town planning in Frontier America," Columbus, Mo., University of Missouri Press, 1980.

¹³ I interpret here what was studied in great detail by Dr. Jamel A. Akbar in his "Responsibility and the Traditional Muslim Built Environment". Ph.D. Thesis, MIT, 1984. Particularly chapter 3, Formation of Towns and Original Growth makes a carefully researched case for the territorial origin of these towns. For all my references to the Muslim vernacular tradition I am in debt to his work.

¹⁴ N.John Habraken. Transformations of the Site, part 3, chapter 2, "Lives of Systems".

¹⁵ It is remarkable that in architectural education no more than in architectural research, the issue of distribution and coordination of design tasks is seldom if ever a subject of study.

¹⁶ Among those discussing designing, the idea of the design process as iterative and cyclic has been generally accepted for some time. I first proposed the form given here in a more primitive version for a professional audience back in the sixties and have used since in teaching to make the point that designing has a 'rational' and an 'intuitive' side. It is only when we regard the design process as one of reaching agreement about forms and their values and as one of negotiation to reach such agreements that a more sustained rationale can be developed from the model. Explicit formulation of constraints becomes a virtue when we must coordinate and agree with others. There is no need for much explicitness when designing is primarily seen as an internal creative process only. It is ironic that only now that we all want to talk to the computer the need for explicitness becomes more generally acute. But in that case, unless we already have a theory about designing as an inclusive process, explicitness may begin to dominate and distort designing instead of helping it.

¹⁷ The diagrams of fig.2.4 are, of course, only approximations of the situation. We are talking about classes of forms distinguished by properties (constraints. I suspect a 'solution-space' may not necessarily be adequately represented by a continuous surface.

[18] A more extensive discussion of the idea of thematic systems can be found in my book "Transformations of the Site". This book is not about designing but about understanding the built environment as such, regardless of the question whether it comes about by designing or not. Of the various ways in which order comes to the built environment thematic systems are only one. Theirs is the only order that is purely the result of human preference. Where physical properties like gravity, size, and direction of movement, bring their own order as do the territorial instincts that we share with the animal world, there is a good deal of freedom of choice left. It is in this free space that we make rules and accept limitations for purely social and personal reasons.

In the present essay, however, we look at artifacts as products of our preferences; that is to say: variants in thematic systems and, from that perspective, encounter physical and territorial constraints. This is the way of looking at the world that we are most familiar with.

[19] It is in the idea of agreement behind the values of the built environment that the links and disconnections between what I seek to formulate and Christopher Alexander's approach may become more clear. To me a pattern is the result of a social agreement first of all. Alexander's pattern book is a rich storehouse of form references with often attractive arguments why they are good. But the true worth of a pattern, it seems to me, is that in each case again those who are involved must agree to accept it. Not on the basis of an outside authority or research, but on grounds of the arguments among themselves and on the basis of their experiences. The pattern is most valuable as the result of implicit agreement and shared experience. The 'good' pattern is the one that reflects consensus; the best one is self-evident and never discussed. Pattern are, indeed, a result, not a means. Patterns that reflect such agreements, implicit or explicit, are what thematic systems and types are made of as well.

[20] The view of the artifact as a machine is most admirably presented by Herbert A. Simon in his "The Sciences of the Artificial." (MIT press, first published in 1970) Because this book has so many qualities and because Simon is willing to think boldly and broadly and is not beholden to any parochial professional ideology, it is the best representation of the machine view that I try to distinguish here.

Simon has only one sentence to say about environmental design and that, as is characteristic for him, is quite to the point and fresh for the time he said it: "We have usually thought of city planning as a means whereby the planner's creative activity could build a system that would satisfy the needs of the populace. Perhaps we should think of city planning as a valuable creative activity in which many members of a community can have the opportunity of participating - if we have the wits to organize the process that way." (page 75)

Simon cannot pursue this line of thought, however, for two reasons. Firstly, the concept of town or city does not invite seeing it as an object with a 'behavior' in an 'environment' which is what he is interested in. He implicitly holds the notion that we are always outside the artifact we are dealing with. From this position one can also view people - "a man, viewed as a behaving system, is quite simple...." (page 52) but not houses or towns.

Secondly, in city planning, Simon suggests, others must participate. But this insightful comment he already sets aside in the context of his general argument. It is not only in city planning that we communicate with others, negotiate, must come to agreements and make those appearances that are acceptable within the social body in which the artifact is made and in which it must exist. This need for interaction among people is the very basis of all designing (that is to say; of the design act, not the product.), both in architecture and in engineering. It is not enough to assume that professional knowhow and goodwill can, by itself, take care of the emergence of artifacts.

Whereas Simon omits the social dimension of designing, he pays attention to the designer's psyche but does not make the distinction between designing and creating that seems to me essential.   Where all making needs creative power, designing, as I have argued, is a special way of being creative. His book is therefore partly a treatise about what the creative mind does, which brings him into psychology and cognitive theory, and partly about designing machine-like things. Both are extremely interesting subject but together they only partly overlap with the subject of designing.

21 Hierarchy is, of course, a familiar means of dealing with complexity. Here again Simon has valuable things to say in the fourth part of The Sciences of the Artificial. My attempts to contribute to this subject can be found in Transformations of the Site where I argue that our exercise of control over the built environment creates the particular hierarchies we find in it. In the context of the essay here I reiterate the argument to make the point that hierarchies in the form that we make, although artifacts themselves, also frame human action. The comment in chapter 11 of this essay, that levels in form hierarchies are domains of intervention introduces the subject of the fourth essay on control of form.

22 The idea of control is, to my mind, an important element in our understanding of designing. But it is, in our inquiry here, only one aspect. It is in my books The Structure of the Ordinary and the earlier Transformations of the Site that control, implying the exercise of power and the resulting changes of form, is the central concept.

23 This chapter relies on what I have learned from Jamel A. Akbar's study: "Crisis in the Built Environment, The case of the Muslim City." E.J.Brill, New York, 1988, IBN 90-04-08758-3

24 In contemporary practice architects are known to claim ownership of the form's design, arguing the building to be an artwork. Cases are known where an owner was forbidden to transform a building in a court judgement on those grounds.

25 In contemporary practice architects are known to claim ownership of the form's design, arguing the building to be an artwork. Cases are known where an owner was forbidden to transform a building in a court judgement on those grounds.

26 For a more extensive treatment of the concept of territory in environmental forms, see my book The Structure of the Ordinary: Form and control in the built environment. MIT Press,1998.

27 For a demonstration of the uses of both territorial and systemic division of design tasks for the design of an urban environment by a team of designers see: The Grunsfeld Variations, a Report on the Thematic Development of an Urban Tissue. N.John Habraken with J.A. Aldrete, R. Chow, T. Hille, P. Krugmeier, M. Lampkin, A. Mallows, A. Mignucci, Y. Takase, K. Weller, and T. Yokouchi, Laboratory of Architecture and Planning, MIT.

28 Fig.4.6 given in this chapter is from the paper by C. Niessen, Hierarchical Design Methodologies and Tools for VLSI Chips, Proceedings of the IEEE, January 1983.

29 "Very Large-Scale Integration (VLSI) will soon make it economically viable to place 1 000 000 devices on a single chip...." Carlo H. Sequin, in Managing VLSI Complexity; an outlook. in Proceedings of the IEEE, Januariy1983.

30 As is typical for a young and dynamic field, terminology in VLSI design is not uniform nor always clearly defined, particularly because a theory of hierarchic structures in VLSI layout design is only beginning to unfold. The naming of levels given here I got from my correspondence with Ir. A.J. Huart, Philips Gloeilampen Fabrieken, Eindhoven. For the little I know about VLSI layout design I am much indebted to him as well as to Ir. C. Niessen and Ir. H. Bosma of Philips Research Laboratories in Eindhoven, the Netherlands.

31 The first time that I saw the principle of window flexibility made available to the users of rental units was in the experimental housing project known as "Molenvliet" in the municipality of Papendrecht, The Netherlands, designed by architect Frans van der Werf, in 1974.

***

THE APPEARANCE OF THE FORM.

N. J. HABRAKEN

For Product Safety Concerns and Information please contact our EU representative GPSR@taylorandfrancis.com
Taylor & Francis Verlag GmbH, Kaufingerstraße 24, 80331 München, Germany

www.ingramcontent.com/pod-product-compliance
Lightning Source LLC
Chambersburg PA
CBHW050521300426
44113CB00026B/1340